21st Century Atlas of the Moon

21st Century
Atlas of the Moon

Charles A. Wood &
Maurice J. S. Collins

Morgantown 2013

Authors

Charles A. Wood
Wheeling, WV 26003
USA

Maurice J.S. Collins
Palmerston North, 4414
New Zealand

West Virginia University Press 26506
© 2013 Charles Wood & Maurice Collins
All rights reserved
First printing 2012 Lunar Publishing, UIAI, Inc., Wheeling WV 26003

 20 19 18 17 16 15 14 13   9 8 7 6 5 4 3 2 1

ISBN:
PB 978-1-938228-80-3

Library of Congress Cataloging in Publication Data
IN PROCESS

Table of Contents

Acknowledgements

We thank the Lunar Reconnaissance Orbiter Camera team for designing and operating a superb instrument that has brought our knowledge of the Moon well into the 21st century.

Additionally, we acknowledge permissions given by our families to take time away from them to compile this atlas. Without the support from Vera, Lilia and Morgan, and Lesley and Shannen, the atlas would not be worth doing.

Introduction

The Moon is the most fascinating object in the night sky, and although less dramatic, it is often unobtrusively present during the day. With a small telescope tens of thousands of features can be observed, some as small as a few kilometers across. Nowhere else in the universe can you see the nooks and crannies of another world. Nowhere else can you always observe changes due to shadows moving across a topographically exuberant landscape. The Moon can enthrall you for a lifetime. The goal of this atlas is to help you find your way among its craters, clefts, volcanoes and mountains, and suggest features that are most appealing to search out, and why.

Full Moon 2012 May 5, 8-inch SCT. Maurice Collins

This guide and atlas will be useful especially to amateur astronomers while observing with their backyard telescopes, or when indoors when they want to identify and learn about features. It will even aid professional scientists who would like to gain a personal knowledge of the world they study remotely. Hopefully, the most important readers will be the inquisitive public. These are adults and children who are alert to the world they live in, who pick up shells at the sea shore, notice different

Artist concept of LRO in lunar orbit. Chris Meaney/NASA.

types of clouds, can name multiple species of trees, flowers and birds, and who count the seconds following a flash of lightning. To these people the natural world is alive and full of fascination. We want to enlarge their realm of enquiry and awe to reach the Moon. It is always there, as it has been for 4.5 billion years, and it never ceases to pleasantly surprise us whenever we unexpectedly see it. We hope it will enchant you too.

The Lunar Reconnaissance Orbiter, still circling the Moon as we write, has collected a vast repository of high resolution and consistent illumination images of our Moon. LRO scientists have mosaicked thousands of the Wide Angle Camera 100 m/pixel images into global maps, and we are pleased to have received encouragement from the LRO Mission Scientist, NASA's Dr. Richard Vondrak, to create this atlas.

One Beginning, Two Different Worlds

The formation of the planets of our solar system is thought to have occurred by the collisions of small rocky masses, called planetesimals, that condensed out of a rotating cloud of dust and gas. Collisions caused some planetesimals to accrete or grow rapidly and become the ancestors of the planets. During the late stage of accretion it is postulated that the proto-Earth collided with a somewhat smaller planetesimal, and that the resulting debris thrown into orbit around the Earth re-accreted to form the Moon. There is much evidence from the Apollo lunar missions and elsewhere to support this apparently bizarre notion, but there are also some observations still to be explained by the theory.

The Moon is a much simpler world than Earth. It is smaller and more quickly radiated away the heat of its energetic formation and from early radioactive decay. Its small mass made it impossible to gravitationally retain an atmosphere, so even if one formed from impacts of water-rich comets the gases quickly escaped to space. Without the erosive effects of air, water or plate tectonics the Moon doesn't change very fast. In fact, most of its surface is much older than nearly anywhere on Earth.

Both Moon and Earth were heavily cratered from the sweep-up of debris left over from the formation of the solar system. The heavily cratered areas of the Moon such as visible near the lunar southern hemisphere are parts of this ancient crust. Radioactive decay of potassium, uranium and other elements caused massive mantle melting and vast outpouring of lava across the surfaces of both worlds. On the Moon those lavas are the dark patches visible with your naked eye, and they are typically 3.5 to 2.5 billion years old. The only conspicuous things younger than that on the Moon are a small number of large impact craters. For example, one of the very youngest is Tycho, which formed about 109 million years ago.

Earth preserves almost none of its surface from the time of lava flows on the Moon. Only tiny pieces of the Earth's surface rocks date from 3.8 billion years ago, and the average age of our world's surface rocks is only 0.5 billion years. When youthful Tycho formed on the Moon dinosaurs roamed the

Earth. Plate tectonics continually recycle Earth's rocks, carrying old material down into the mantle, and volcanically creating new rocks at ocean floor spreading centers and along continental margins. Earth is full of heat that drives plate tectonics and dynamically reconfigures its surface over millions of years. The Moon is dead except for small scale fracturing and the pinging of its surface by a rain of tiny impacts, and every few tens of millions of years one big enough to create a crater visible from Earth.

The Earth – Moon pair is the most important collection of objects in the solar system, not just because we live on one of the worlds (and hopefully someday on both), but because the Moon preserves what most objects in the early solar system must have looked like, telling us of the great importance of impact cratering. And Earth is the most evolved planet in the solar system, showing how an inner heat dynamo transforms a world beyond its simple beginning. If we lived on Mercury, Venus or Mars we would have a much less rich understanding of the history and development of planets and the solar system.

What You Will See on the Moon

Whether you look at the Moon through a telescope or examine images of it you will see a vast, almost bewildering number of landforms. The good news is that there are really only three different types of features, and all the myriads of lunar structures are simply large or small, young or old varieties of them. The three types are impact craters, volcanic lava flows and eruption sites, and fractures of the lunar crust. Or to put it in terms of geologic processes, impact, volcanism and tectonism. Missing are all the other processes familiar from Earth: the water cycle and its features (such as rivers, seas and glacial landforms), sand dunes and other wind-blown deposits, and all of the types of fracturing and volcanism associated with plate tectonics, the horizontal and vertical moving of the crust and upper mantle of the Earth. And of course the effects of life.

Impact Craters

Impact craters result from the collision of a comet or asteroid with a planetary surface. These projectiles may be only microns across or up to tens of kilometers wide; with large ones being much rarer than small ones. All of the projectiles orbit the Sun with velocities ranging from a few kilometers/second to 70 km/s, and the energy of an impact depends on the square of the velocity. So first of all, the velocities are much higher than we normally encounter on Earth, and secondly, squaring them makes the energies of impact far beyond any Earthly experience. Consequently the craters produced are very different than what we see on Earth. Earth craters would initially look similar to lunar craters, but big craters occur so infrequently – the crater-forming impact that killed the dinosaurs happened 65 million years ago – and Earth's erosional processes are so fast, that nearly all of the 250 identified impact craters on Earth are highly degraded compared to their lunar cousins.

Simple Craters

Freshly formed small lunar craters – ones with diameters smaller than about 15 km – are perfect bowls with steep walls and small flat floors; they look like they were turned out on a lathe. The high-speed impact smashed through the target rocks and in a sense exploded, throwing debris out of the growing hole. Once everything quieted down the final

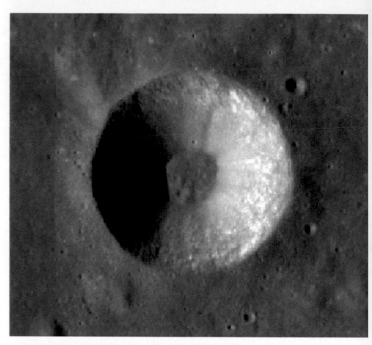

13 km wide simple crater Chladni. LRO/NASA

craters had depths about 30% of their diameters, The hyper-resolution images from the Lunar Reconnaissance Orbiter Narrow Angle Camera reveal that boulders and impact melt are everywhere on the walls and floors of these so-called simple craters, but what we see from Earth conceals that micro scale complexity. Simple craters are simple.

Complex Craters

More energetic impacts create larger, more interesting craters, called complex craters. Because of the immense energy associated with bigger impacts the ground zero point gets strongly compressed and finally after a massive pressure wave passes, the ground decompresses and previously

83 km wide complex crater Tycho. LRO/NASA

deeply buried rock units rebound up in the center of the crater floor. That is how central peaks form – and because their rocks come from great depths they are important in determining the composition of the lunar crust.

The formation of large craters excavates such a big and steep-walled hole that the surrounding rocks collapse into the hole. You have seen the results when looking at craters like Tycho and Copernicus – their inner walls are wreathed in concentric terraces of rocks that slid down-slope. Smaller complex craters – typically ones 15 to 40 km in diameter – have walls that fail on only one or two places, so instead of massive, near continuous terraces they have individual slumps of debris on their floors, and a notch in the rim where the collapse occurred. The 24 km wide Lalande is a great example, with a massive landslide below the alcove in its western wall.

Each impact event excavated a hole and ejected vast amounts of material – that is why we see craters. Some ejecta landed near the impact point, constructing the elevated rims that surround central depths. Much of the ejecta was deposited beyond the rim, thinning with distance. Nearby craters are often strongly modified with infill and their rims are softened. Steamers of ejecta, called rays, were hurled hundreds to a thousand kilometers away from the craters, draping the surface with bright, pulverized ejecta. Boulders entrained within the streamers excavated small secondary craters when they landed, creating more splashes of brightness. Over time the pulverized bright material darkens because of the continuing bombardment of the surface by particles from the Sun. Without this space weathering every crater ever formed would be the source of radiat-

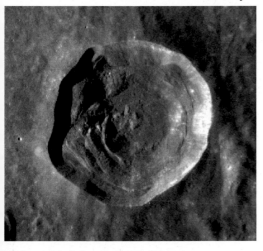
24 km wide Lalande. LRO/NASA

ing rays, and the Moon would be crisscrossed, as if wrapped in spider webs. Relatively young complex craters like Lalande and Tycho have depths only about 10% of their widths. Complex impact craters have cross-sectional shapes like saucers, and simple craters are more like the tea cups that sit in them.

Basins

The most energetic impact events created craters that are so big and so shallow that they were not noticed for hundreds of years. Part of the reason for missing them is that the youngest and most dramatic one is on the western limb of the Moon, and its full structure could not be appreciated until Moon-orbiting spacecraft photographed it from overhead. Such giant craters are so different from smaller ones that they have a different name – impact basins.

The Orientale Basin is named for the small patch of dark lava that was glimpsed from Earth and named Mare Orientale. You are probably scratching your head – why was a feature on the Moon's western limb named Orientale, for the word Orientale means eastern?

1100 km wide Multi-ringed Orientale Basin. LRO/NASA

History provides the answer. For hundreds of years the Crisium side of the Moon was considered the western limb and Orientale was the eastern limb. In 1960 the International Astronomical Union, the organization that sets rules for planetary names – more about them later - flipped the east and west directions to make them compatible with how the terms are used on Earth – east is the direction of rotation of the Earth and now the Moon. Every observer in 1960 was outraged but they are mostly dead now and all the newcomers never knew it was different.

Back to the basin misnamed Orientale. Images from orbit show a series of concentric mountain ranges surrounding the central lavas. The outer and most dramatic is the Cordillera Mountains. From Earth, observers saw them in profile against the limb, and the orbiter image shows that they continue around the entire basin. Inside are two more, lower rims called the Inner and Outer Rook Mountains. Measurements show that the floor of Mare Orientale is about 11 km below the rim crest of the Cordilleras. This is twice as deep as the deepest complex crater, but when compared with its 950 km diameter, the Orientale Basin is nearly flat; the depth is only 1% of its width.

The beautiful circular symmetry of Orientale confirmed what a few pioneers had proposed earlier. Imbrium, Nectaris and other circular patches of mare lava were inside huge multiringed basins. But none of these Earth-facing basins are as pristine as Orientale so that their circular and multi-ring structures were not as obvious.

One thing that had been noticed was that many of the patches of lava, called by the Latin word mare (for they reminded 17th century observers of Earth's seas) were circular. Mare

600 km long Apennine Mountains arc of Imbrium Basin rim. *Consolidated Lunar Atlas*

Crisium near the east limb is a great example for it is nearly completely surrounded by a mountainous rim. It is mostly circular but doesn't look like it because it is foreshortened into an oval from our earthly observing point.

Most circular maria (plural of mare) are at least partially surrounded by mountain ranges. The two most spectacular and easily visible ones are the Apennines and the Altai. The Apennines and their less spectacular continuation, the Carpathians, define about 30% of the circular rim that may have originally surrounded the Imbrium Basin. The northern rim of the basin is not well defined and the western rim is missing altogether, as if the basin were tilted downward to the west. The later flooding of the Imbrium Basin by massive sheets of lava, creating the mare, may have overtopped a western rim, and it certainly covered most inner rings. At Orientale the inner rings and basin floor are well shown because floods of mare lava did not inundate that basin.

Isolated peaks and ridges in Mare Imbrium seem to define two circles that could represent inner basin rings. Peaks such as Pico, Piton and La Hire may be the high points of otherwise lava-covered circular mountain ranges.

The Atlai Scarp dramatically bounds the western edge of the Nectaris Basin, named for the small Mare Nectaris in its center. Because there is much less lava fill in Nectaris, its two inner rings, despite being low and battered, are easily seen once pointed out (see page B4 for an overhead view). A third circular mare contained within an impact basin is Mare Humorum. Here the surrounding rim is lower and less conspicuous but it is detectable once the pattern has been observed at other basins. Mare Serenitatis fills another apparently older basin, and the Haemus Mountains are almost all that is left of its original rim.

When a giant impact excavated a basin-size hole hundreds of kilometers wide, an immense quantity of rocks, boulders and dust was thrown beyond the hole. Some of it traveled in ballistic trajectories across the lunar sky and was deposited

up to a thousand or so kilometers beyond the basin. Many pieces of ejecta undoubtedly travelled so rapidly that they escaped from the Moon and hit the Earth and other planets and perhaps went into orbits around the Sun. Brilliant rays must have emanated from basins, but because all basins are more than 3 billion years (b.y.) old, space weathering and subsequent impact cratering has long since erased them. But secondary craters formed in association with rays remain, and because impact basins were far more energetic than smaller craters, their ejected material was larger and the secondary craters are too. The best seen from Earth is the Rheita Valley (Chart 5), a 450 km long line of closely spaced secondary craters, each 20-30 km wide. But far better preserved basin crater chains are out of sight around the western limb beyond Orientale.

Once again, the relative youthfulness of the Orientale Basin means that its ejecta are best preserved. However, much of its eject that came to the Earth-facing lunar hemisphere was covered by subsequent mare lavas. The farside deposits of Orientale ejecta are undisturbed and we can see that they are much more extensive and continuous. Thousands of large and small craters were completely buried by Orientale ejecta, others had their floors filled in by it, and often their walls were destroyed too. The best place on the nearside to see the effects of instant modification by basin ejecta, in this case from Imbrium, is the region around Julius Caesar (Charts 11 & 12).

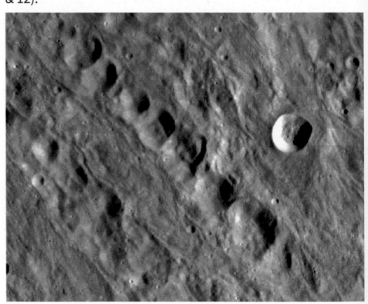

Leuschner Crater Chain, farside. What the Rheita Valley may have looked like soon after its formation. LRO/NASA.

Impact basins are the largest structural elements of the Moon. They caused the lowest – their floors – and highest – their rims – lunar topography, and deep fractures under them provided conduits for magma from the mantle to rise to the surface forming the maria and small scale volcanic landforms. Basins determine the architecture of the lunar surface.

Volcanism

About 17% of the lunar nearside is covered with mare lavas, and since maria are much rarer on the farside, probably less than 10% of the entire lunar surface is volcanic rock. For comparison, oceans cover 70% of Earth, and the rock under the water is volcanic. Mars and Venus also have a much greater amount of surface volcanism than the Moon. Small worlds such as the Moon lost their heat earlier than more massive planets so that their duration of volcanism was shorter. The Moon's main period of volcanic eruptions was between about 3.5 and 2.5 billion years ago, and only small amounts of lava dribbled out afterwards. The most recent eruptions are thought to date from about one billion years ago. During the last two centuries there have been hundreds of observations of clouds, glows and obscurations that often have been enthusiastically interpreted as eruptions; almost certainly they were not. The Moon may have a small amount of molten material in its core but the lunar surface has been volcanically dead for at least a billion years. R.I.P.

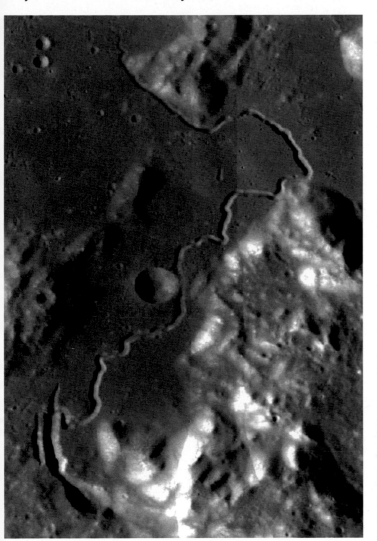

Hadley Rille width is about 1.2 km. LRO/NASA.

When the Moon was eruptive it must have been a magnificent sight. Mare lavas were much more fluid than terrestrial ones and flowed across basin floors like honey rather than molten rock. That explains how individual lavas – as seen in western Mare Imbrium (Chart 21) – could have flowed hundreds of kilometers even on gentle slopes. Tens of thousands of flows were required to build up the thick piles of lava that make up the maria. For example, the Imbrium lava pile is estimated to be about 5 km thick in the center of the basin, and apparently the lavas erupted over a period of at least one billion years.

Very few vents - places where lavas erupted onto the surface - have been preserved. But fortunately many of the channels and tubes that fluid lavas typically create, as witnessed in Hawaii, still exist. These sinuous rilles are usually only a few

kilometers wide, and a hundred meters or so deep, and yet they are hundreds of kilometers long. The Apollo 15 mission made a pinpoint landing near the Hadley Rille (Chart 11), making it the most visited sinuous rille on the Moon. Some sinuous rilles start in a short curved trough that appears to be a collapse over the vent. A number of interestingly shaped vent depressions occur at the start of the Prinz Rilles (chart 21).

Some mare lavas were not as fluid or perhaps did not flow as rapidly as normal lavas and built up small mounds around their vents. Typically, these domes of lava are 5 to 15 km in diameter and only a few hundred meters high, resulting in slopes that are so gentle that the domes can only be detected when the illumination is low. A large relatively easy to spot example is the Kies Dome (Chart 23) just west of its namesake crater. A famous cluster of six domes occurs just north of Hortensius (Chart 22), and like the Kies Dome many of these have volcanic vent pits at their summits. A very few eruptions had either very viscous lavas or perhaps included ash so that they constructed small cones with wide craters. The largest – but still only a few kilometers across – is Marian T (Chart 20).

Explosive volcanism is commonplace on Earth, but rare on the Moon. The reason is that gas is needed to expand within magma, ultimately inflating it so much that it shreds the solid rock into small particles of ash, or pyroclastics, as volcanologists call them. On Earth, water is the major gas that causes explosive eruptions. The extreme dryness of the Moon, despite recent identification of ice in polar craters, means that some other gas must play water's role if lunar pyroclastics exist. And they do. Green and orange spheres of volcanic glass produced in pyroclastic eruptions were discovered at the Apollo 15 and 17 sites. These glass beads were produced in gas-rich eruptions that deposited dark material, visible from Earth, in those regions and elsewhere.

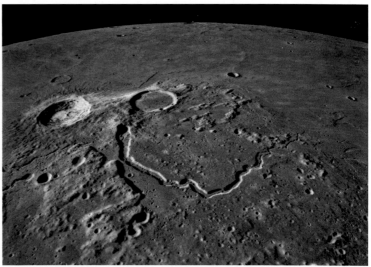

Aristarchus Plateau, the bright crater Aristarchus is 40 km in diameter. Apollo 15.

The largest pyroclastic unit on the Moon drapes the Aristarchus Plateau (Chart 28). The ashes erupted from the Cobra Head, a large volcanic vent for Schröter's Valley, by far the largest sinuous rille on the Moon. This pyroclastic deposit erupted before the lavas that surround the Plateau because the sharp boundary at the plateau's edge shows that the mare lavas cover the ash that must have fallen there. The Aristarchus Plateau has a very faint color that has been reported as a pale mustard tint or a faint green hue. It is the most conspicuous color that an eyeball can sometimes detect on the otherwise monochrome Moon – at least when seen from 384,000 km away.

Sometimes only a small amount of pyroclastic material erupts. This is the situation for dark halo craters or DHC. The most famous DHC are a cluster along the eastern floor of Alphonsus (Chart 16). About a half dozen rimless pits 2-3 km in diameter occur along thin rilles. The pits are surrounded by dark ash halos that are two or three times larger than the pits; they are easy to see under full Moon conditions. Warning: other dark halo craters, such as Copernicus H (Chart 17), are impact craters that excavated through bright ray material to bring up underlying dark mare rocks. All volcanic DHC occur inside larger craters of the floor-fractured type. DHC not inside craters are almost always impact craters, not volcanoes.

Volcanic dark halo craters and rilles on the floor of 111 km wide Alphonsus. LRO/NASA

What are floor-fractured craters and why are they in a section on volcanic features? Alphonsus is a floor-fractured crater – FFC - and so are many of the most interesting large craters on the Moon. A FFC is an impact crater typically larger than 40 km in diameter that formed near the edge of a basin. When magma rose up basin fractures to form maria, some ponded under nearby craters, and if the pressure in the magma increased sufficiently the floor of the overlying impact crater was lifted, central peak and all, creating a ring of fractures . Commonly, these fractures along the edges of the floors are marked by rilles and/or ridges, and sometimes lava erupted onto the floor making a DHC, sinuous rille, or pond of lava. Posidonius (Chart 8) is one of the most remarkable FFCs; apparently its uplifted floor was also tilted. Petavius and Humboldt (both on Chart 4) are other dramatic examples with multiple volcanic features.

Lunar volcanic rocks are rather monotonous chemically, compared to the wide varieties of those from Earth. Nearly all the volcanic features describe above are made of mare lavas, which do have differences in the amounts of iron and titanium and a few other elements – accounting for why some mare lavas are dark and other lighter. But a few places on the

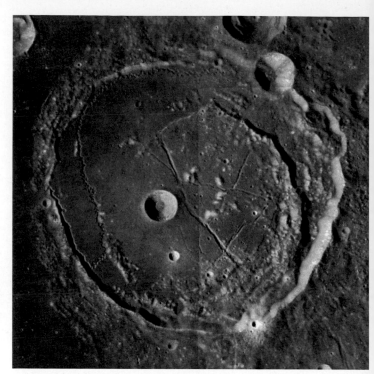

101 km wide floor-fractured crater Posidonius. LRO/NASA

Moon also have rock compositions that seem to be completely different from what is normal. This non-mare volcanism manifests itself as having spectral colors (detected by careful imaging through filters) that are brighter in the red end of the spectrum than are mare lavas. Perhaps the most informative of these red spot lavas are the two stubby mountains named Gruithuisen Delta and Gamma (Chart 21). These are steep-sided domes made of fairly viscous lavas that piled up around their vents instead of flowing away. Another red spot is the 30 km wide Arrowhead, more prosaically called Mount Hansteen (Chart 26). Here the volcanism built knobby hills that an instrument on the LRO spacecraft identified as one of the most silica-rich areas of the Moon (the Gruithuisen Domes are similar). What this means is that although lunar volcanism is overwhelmingly of the fluid mare type, in some locations granite-like magma erupted to the surface. Someday red spots will be targets for detailed on-the-ground investigations.

Tectonics

Tectonics is the word that describes the movements of the crust of a world. On Earth, plate tectonics drives the dynamic changes of the planet, causing eruptions and earthquakes, and breakup and convergence of land-masses. The Moon has never had plate tectonics; impact cratering and volcanism cause all of its crustal fractures.

As we saw earlier, complex craters are large enough that the surrounding terrain collapses into the newly created holes, forming terraces. Each scarp of a terrace is a fracture and the rims of craters are tectonic features. A very similar but larger scale process occurs around multi-ring basins. The Apennines and Altai rims are giant fault scarps, where the inner parts of the Imbrium and Nectaris basins subsided downward. Basin rims like this are the largest faults on the Moon.

The one to five kilometer thick piles of lavas that fill a basin create an immense load, and because the lavas come from beneath the basin, reduce its support. As basins fill up with lava they subside, causing fracturing around their edges. Most circular basins are surrounded at least partially by

Earth. The Wall is about 120 km long with a height of about 400 m.

North end of Sirsalis Rille. LRO/NASA

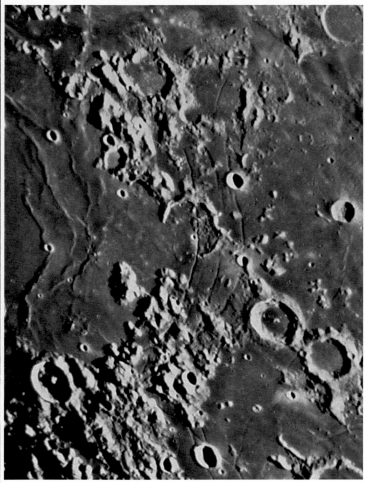

The 57 km wide crater Hippalus is cut by arcuate rilles at the hinge line of Mare Humorum – at left. CLA.

rilles that are subsidence cracks. The best example is the set of flat-floored rilles that curve around the eastern side of the Humorum Basin (Chart 23). Narrower concentric rilles arc around the northern and eastern shores of Serenitatis (Charts 11 & 8) and around the western and southern edges of Tranquillitatis (Chart 12).

Another type of flat-floored rille occurs as linear troughs. The Hesiodus Rille (Chart 16) cuts the southern side of Nubium but doesn't seem to be a basin-concentric rille. A number of linear rilles approximately radial to the edge of Ocean Procellarum suggest that they are related to forces associated with that giant structure, The biggest of these approximately radial features is the Sirsalis Rille (Chart 26), which is a 400 km long trough. Surprisingly, the strongest magnetism of the Moon is along this rille.

This encouraged the interpretation of Sirsalis and similar linear rilles as surface traces of vertical sheets of magma (called dikes by geologists) that conveyed magma to Oceanus Procellarum. In fact, a few linear rilles have very small cones and pyroclastic splotches that show where magma from dikes erupted onto the surface.

Finally, we come to what most observers think of when they see the word fault. The Straight Wall (Chart 16) may be the most famous fault in the solar system other than those on

When the Sun is shining from the east, the Wall's westward shadow suggests a steep slope but it is only about 21°, a steep climb, but not a vertical wall. The Wall spans an old crater sometimes called Ancient Thebit. The entire floor of this 200 km wide ruin is covered by lava, and the western half of its rim is missing, although low mare ridges are ghostly traces of where it used to be. It appears that Ancient Thebit was cut by the Straight Wall fault, which lowered the western half enough for later lavas to cover the crater's rim.

Straight Wall casting a shadow. CLA

Although no fault scarp is visible there, it seems likely that the same thing happened to the original Sinus Iridum crater (Chart 20), submerging the down-dropped southern rim below the subsequent Imbrium lavas.

Two more major faults are recognized on the Moon and a number of smaller ones occur. The 115 km long Cauchy Fault in western Mare Tranquillitatis Chart 7) is mirrored by a similar length rille – both are about radial to the Imbrium Basin and somehow must be related to the giant collision that formed it. This region is also full of domes.

The Lacus Mortis Fault (Chart 10) is different from the other two in tapering from a maximum height of about 400 m to zero, 40 km later where it turns into a rille. Lacus Mortis is an old crater or perhaps a small basin and it is nearly filled with old lava that has a few rilles as well as the tapering fault.

There is one remaining type of lunar fault that is very common - dozens of them even have names - but whose origin observers often do not realize. Mare ridges, which used to be

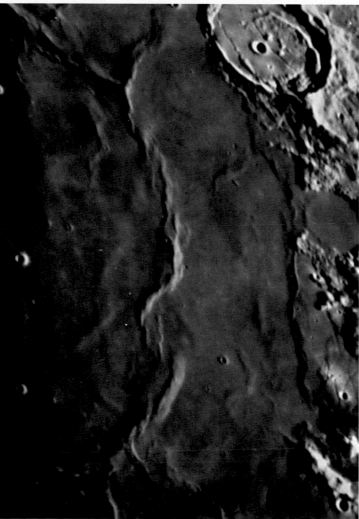

Serpentine Ridge in eastern Mare Serenitatis, with Posidonius at top right. CLA

called wrinkle ridges, are low ridges only a few dozens to a hundred or so meters high, that snake along maria for hundreds of kilometers. The most famous example is the Serpentine Ridge in eastern Mare Serenitatis (Chart 8).

Mare ridges like the Serpentine example are low angle faults that deformed the thick piles of mare lavas. After the lavas

cooled and solidified, the basins that held them subsided due to the weight of the maria, fracturing and forming concentric rilles around their edges. A different type of accommodation was necessary for the massively heavy, conical pile of mare lava to fit into a smaller volume of the sunken basin floor. This subsidence caused parts of the brittle lava to fracture and slide over adjacent pieces. The mare pile became doubled along these ridges to accommodate its reduced diameter. Many concentric mare ridges are along the same circles as isolated peaks that mark buried inner basin rings, suggesting that the buried mountains anchored the faulting.

In viewing the Moon, either at the telescope or by looking at images, a complex story involving cratering, volcanism and tectonics is frequently presented. These processes occurred at different times, sometimes more than once, over-printing and sometimes destroying whatever happened earlier. And all of the landforms are likely to be billions of years old and have suffered all sorts of modifications by subsequent cratering, volcanism and tectonism. The challenge – and fun - in looking at such a landscape is trying to figure out what happened and in what sequence. Once you begin to observe with a geologist's eye you will find that the Moon will entertain and challenge you with a lifetime of pleasurable studies.

Lunar Nomenclature

Early map makers named lunar craters and mountains after their patrons and friends. The second person who charted the Moon ignored the names on the first map, and nomenclatural confusion was normal for 250 years. During the first half of the 20th century, the International Astronomical Union established a standard, accepted nomenclature, but then in the 1970s reintroduced confusion and chaos by adding hundreds of new names that violated traditions and norms. By the mid-1970s actual lunar scientists got back in control of nomenclature, but much of the damage was permanent. One example was the expanded use of Latin – a dead language that no space-faring nation uses! – for names of landform types. For example, crater chains were called catena, mare ridges were christened dorsa, and faults became rupes. In this Atlas we use English words, with these new Latin terms banished to official maps.

Among the earliest named features on the Moon were the dark spots. Galileo called them by the Latin word mare (plural maria), which means sea. Galileo doubted they were water, because he saw no Sun glint, but he observed that they were much smoother than the bright and rough highlands. Fifty years later these seas were given names derived from nautical terms or states of mind. Mare Frigoris is a northern lunar sea called Sea of Cold, and nearby Mare Imbrium is the Sea of Rains. More benign are the seas of serenity, tranquility and nectar. The farside's Sea of Moscow may represent a state of mind.

Mountains are named after ones on Earth, which can cause confusion for scientists doing comparative planetology. When you say Apennines or Alps you must hope the context will indentify if Earth's or Luna's are being discussed – in this Atlas it is always the Moon!

Craters are named for scientists and philosophers. Many of the famous Greeks such as Plato, Archimedes and Aristarchus are honored with prominent craters in the northern hemisphere. Scientists from the 16th to 18th centuries – Galileo, Kepler, Newton, Gauss, etc - are commemorated so that studying the Moon recalls history too. On the side of the Moon facing Earth 1,240 craters, mountains and other features carry names, and another 690 named features are on the farside. The Moon has far more craters than this – perhaps 25,000 larger in diameter than a few kilometers, and 7,056 of these have received letter designations. For example, Copernicus A is a small crater near Copernicus. Only a few of the lettered craters, ones that illustrate interesting features, are shown on the charts. For example, Petavius B (Chart 3) is a beautiful low angle impact crater whose unique ray pattern should not be missed when the Moon is full. Whenever a letter is placed on a chart, the letter is on the side of the crater closest to its patronymic crater.

This Atlas shows nearly all of the official names of features on the nearside, but some are purposely excluded. For example, 45 names are given on Chart 2, but another 32 are not. This is an area where Apollo era mapmakers went crazy and added many dozens of names for small and insignificant craters just so their map charts would have names that didn't include lettered craters - this was one of the worst mistakes in lunar nomenclature history. Thus, this atlas is not a comprehensive reference for lunar names - the I.A.U.'s *Gazetteer of Lunar Nomenclature* website does that; this atlas just shows the names that fit this scale map.

The Atlas

Maps for this Atlas were created from the Lunar Reconnaissance Orbiter Spectacular Mosaic captured in December 2010 and published online at: http://wms.lroc.asu.edu/lroc_browse/view/wac_nearside. The mosaic is credited to NASA, the Goddard Space Flight Center, and Arizona State University. We divided the nearside mosaic into map charts using the slice tool in *Adobe Photoshop* software.

The Atlas depicts the Moon with the north polar regions at the top, and the Crisium or east limb at the right. This is the official orientation used in maps and photos since 1960, and this orientation facilitates comparison of Atlas charts with such documents.

For Observers:

The map view is almost never the way the Moon is seen through telescopes. It is the way the Moon appears when observed from Earth's southern hemisphere with a Newtonian telescope when the Moon is high in the sky. But if, like most astronomers on the planet, you live in the northern hemisphere, the Newtonian view of the Moon will be south up with east to the left. Many amateur astronomers actually use Schmidt-Cassegrain or refractor telescopes with a right angle mirror in front of the eyepiece. With such telescopes in the northern hemisphere north is up but east is to the left. Ugh. If you live in the southern hemisphere, south will be up and east will be to the right. What all of this means is that no single map will be good for all observers around the globe with every kind of telescope. Blame the laws of optics and the curvature of our world for making it so difficult to match an eyepiece view to an atlas view. The merit of the official orientation is that it is equally awkward for almost everyone.

The Atlas is divided into 28 charts starting at the northeast limb of the Moon, and moves south along the limb, and then shifts west to the next row, and gradually - like the progressive pattern used to mow a lawn of grass - ends at the Moon's northwest limb. This arrangement places areas to the north or south of each chart on adjacent pages, making it easy for observers to follow the terminator. Numbers of adjacent charts are indicated in the locational mini-map at the bottom of each chart.

All of the charts depict the lunar surface with the Sun shining from the east, and always at nearly the same angle (about 25°) of illumination. LRO scientists selected a Sun angle high enough that the interiors of most craters are visible, but low enough that their eastern rims cast shadows to indicate their topography. Near the equator each chart shows an area of the lunar surface about 700 km long and 650 km wide, and dimensions increase approaching the limbs.

Facing each atlas chart is a page illustrating and briefly describing a handful of geologically interesting features of the area. Following the name of each feature within parentheses – e.g. Cleomedes (B8, 130 km) - is the X,Y position of the object on the map using the letters across the chart top and numbers along the left margin, and the diameter or where appropriate the length, of the object in kilometers. Spelling of feature names and diameters/lengths follow the International Astronomical Union *Gazetteer of Planetary Nomenclature*. Some unofficial names are used – such as the Serpentine Ridge – and they appear in italics on the map charts. At the bottom of each map is a listing of the *Lunar 100* objects appearing on that page – these are great targets to observe.

The images illustrating each described feature come from a variety of sources. Some are from the LRO QuickMap and others are lower Sun LRO Wide Angle Camera (WAC) mosaics specially processed by Maurice Collins. A few images come from the Clementine, Lunar Orbiter IV, and M3 camera of Chandrayaan-1 missions, and a handful are from the 50 year old telescopic *Consolidated Lunar Atlas (CLA)*.

Following the 28 main map charts is a series of specialized maps depicting areas and features not well shown on the main Atlas pages.

Limb Charts

First are 8 pages showing the limb regions. The Atlas shows the Moon at zero libration, the average view seen from Earth. However, at any time of observation the Moon usually tilts one edge inward, bringing limb features into better view and allowing limited observation of a sliver of the farside. Seeing the farside is always a thrill. The limb charts show the limb regions tilted inward 10°, making it easier to see features at the limb and just beyond. These charts were constructed using the LRO Spectacular Mosaic, rotated and illuminated with Jim Mosher's invaluable *Lunar Terminator Visualization Tool* (LTVT).

Full Moon Charts

The next specialized charts are high Sun views, taken from the famous Yerkes Observatory photograph Y842 in the *Consolidated Lunar Atlas*. This image was divided into four quadrants, and a handful of features are identified to aid navigation. Bright material in full Moon views is typically crater rims and rays that newly expose lunar rocks. Over time such bright material is darkened by solar wind that converts iron oxides in rocks into dark blebs of iron metal. The highlands of the Moon are intrinsically brighter than the maria because the highlands have little iron but much aluminum and calcium in the form of the bright mineral plagioclase feldspar. Variation in the hue of lunar dark material commonly reflects the amount of iron and titanium in individual mare lavas. A pronounced boundary occurs along the south shore of Mare Serenitatis with a swath of darker, iron-rich lavas bordering lighter lavas in the center of the mare.

Lunar Basins and Mare Ridges

Impact basins are the most important landforms on the Moon, but their large sizes and commonly degraded ring structures often cause them to be overlooked. This section depicts each major nearside basin in a unique image constructed using the LRO LOLA altimetry in LTVT to create a topographic image with a constant illumination of 1°. This emphasizes low slope features such as mare ridges. These ridges are often inner rings of basins and are structurally important but difficult to appreciate in optical images because Sun angles differ across an entire basin.

Landing Sites

Places where Luna, Ranger, Surveyor, Apollo and other spacecraft reached the Moon are historically and scientifically of prime importance – that is why NASA has recently issued recommendations to preserve landing sites. This section contains larger scale images for all six Apollo landing sites and six others of historic probes; in each image a plus sign (+) marks the touch-down point. Telescopically, no evidence of landings is visible, but these charts identify the locations of prime scientific and historic importance.

Farside

Finally, four quadrant maps from LRO are given as a general reference to the farside. A handful of prominent craters and basins are named for orientation. Scientifically the farside is as important as the nearside, but for all but a handful of humanity the farside is always invisible except for slivers that librations teasingly bring into view around the limbs.

Names

More than 1200 feature names are given in this atlas, and without an index few people could find which chart they are on. Each entry in the index gives the feature name, the map chart/s it is found on, and its diameter or length, taken from newly updated values appearing in the I.A.U. *Gazetteer of Lunar Nomenclature*.

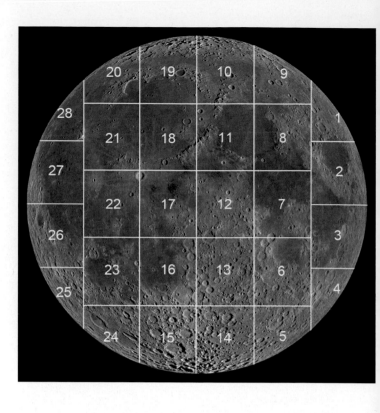

More Information about the Moon

Recommended Publications

- Motamoro Shirao and Charles Wood (2011) *The Kaguya Lunar Atlas*. Springer.
- Paul Spudis (1996) *The Once and Future Moon*. Smithsonian.
- Philip J. Stooke (2007) *International Atlas of Lunar Exploration*. Cambridge.
- Richard Vondrak, John Keller and Chris Russell (editors) (2010) *Lunar Reconnaissance Orbiter Mission*. Springer.
- Charles Wood (2003) *The Modern Moon – A Personal View*. Sky Publishing.
- Charles Wood (2004) *The Lunar 100 Card*. Sky Publishing.
- Charles Wood (1998 onward) Exploring the Moon column in *Sky & Telescope* magazine.

Recommended Online Resources

- Google Moon: http://www.google.com/moon/
- I.A.U. Gazetteer of Lunar Nomenclature: http://planetarynames.wr.usgs.gov/Page/MOON/target
- Lunar Photo of the Day: http://lpod.wikispaces.com/
- Lunar Reconnaissance Orbiter Camera: http://lroc.sese.asu.edu/index.html
- Maurice Collins' Moon Science web site: http://moonscience.yolasite.com/
- Moon Wiki: http://the-moon.wikispaces.com/
- Lunar Terminator Visualization Tool (LTVT): http://ltvt.wikispaces.com/LTVT

The Nearside

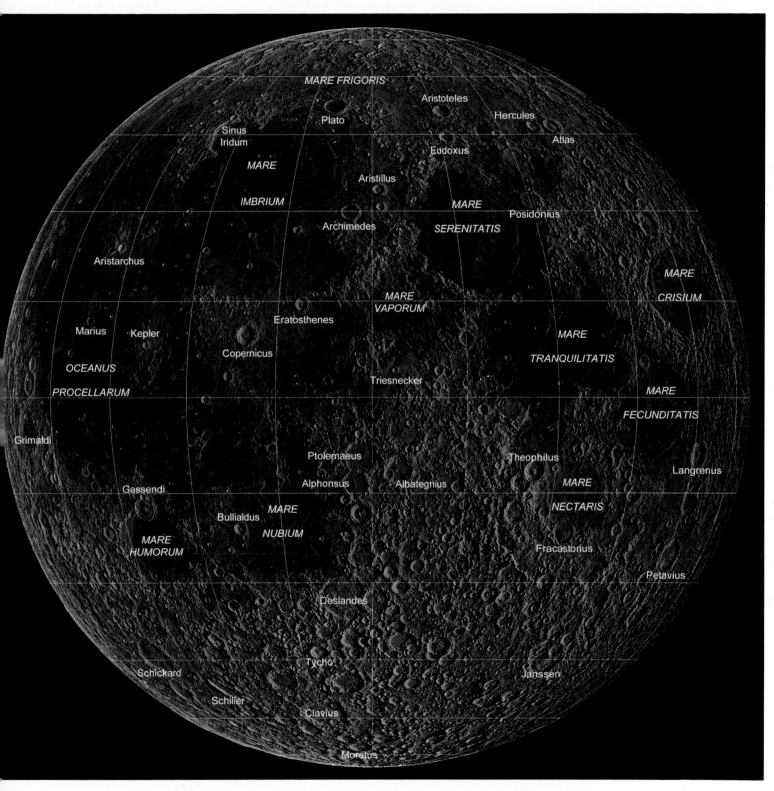

Lunar Reconnaissance Orbiter Mosaic with 15° latitude and longitude lines.

Cleomedes

Lacking large dramatic fresh craters or rille-crossed lava flows the area north of Mare Crisium is populated by three major but degraded craters (Cleomedes, Gauss and Messala), eight to ten medium size ones, and two little-noticed but definitely visible basin rings – the mountainous rim that passes through Eimmart and a scarp from Geminus to Hahn.

Cleomedes (B8, 130 km): Floor-fractured crater with rilles and dark halo craters too small to detect without a large telescope. Tralles smashed onto the northwest rim of Cleomedes, pushing debris onto the big crater's floor.

Geminus (A5, 83 km): The Tycho of the northeast; the same size, but slightly subdued. A complex crater with terraced walls and small peaks, and old enough to have lost its rays.

Gauss (D4, 170 km): One of the largest craters on the lunar nearside, but so close to the limb that its rille-edged interior is unimpressive. The oblique view emphasizes how shallow craters are compared to their diameters.

Messala (A3, 122 km): Large and degraded; shallowed by infill of ejecta from the Humboldtianum Basin?

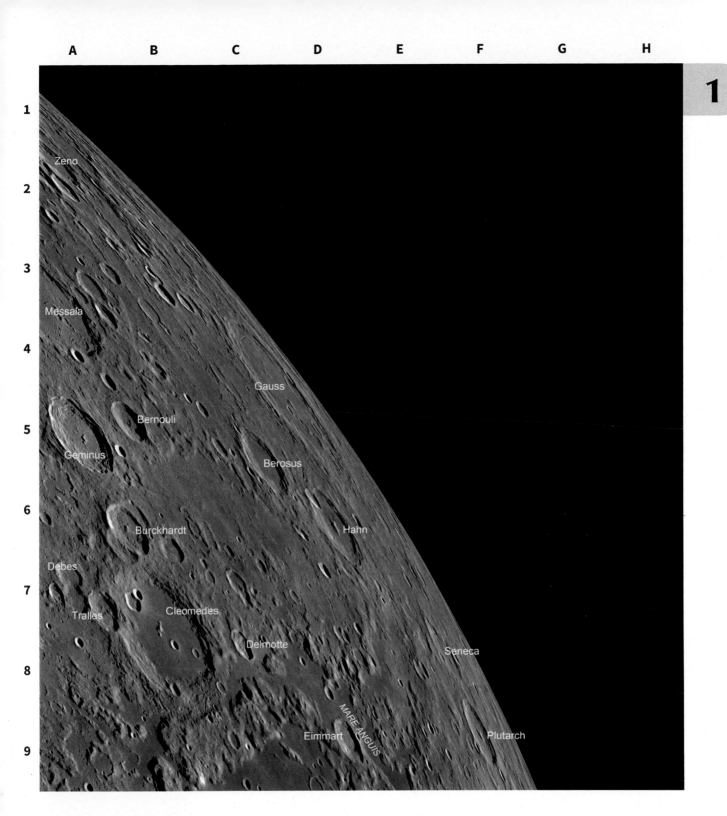

Zeno

Messala

Gauss

Bernouli

Geminus

Berosus

Burckhardt

Hahn

Debes

Tralles

Cleomedes

Delmotte

Seneca

MARE ANGUIS

Eimmart

Plutarch

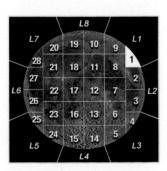

Cleomedes 1

Lunar 100:

None

Mare Crisium is the only maria not in contact with others. It is nearly completely enclosed by mountainous terrain, its basin rim. Actually, the rim is broken near the limb, evidence that the basin was formed by an oblique impact. Another mare – Marginis – that is right on the limb may occupy an ancient basin but there is no compelling evidence. View the entire Crisium and Marginis basins and their mare ridges on Charts B1 and B2.

Mare Marginis (H4, 358 km): Irregular mare with bright swirls just marginally visible with good libration. Antipodal to Orientale Basin (Chart 26). Lava-floored Neper (137 km) at bottom, is often seen nearly in profile with tall bright walls and 2 km high central peak dramatically emerging from shadows.

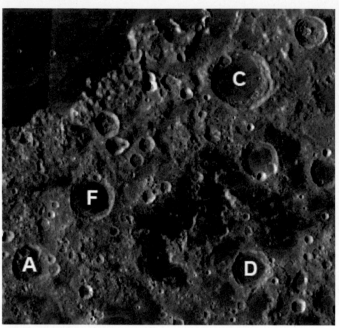

Mare Undarum (G7, 245 km): Mare Undarum is an area of large ponds of mare material filling low spots within the southern continuation of Crisium's Geminus ring (Chart 1). This leaky area includes C = Condorcet, F = Firmicus, A = Apollonius, D = Dubyago, and two unnamed craters with Mare Undarum in the middle.

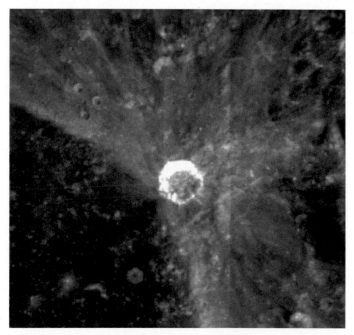

Proclus (A3, 27 km): Brilliant walled oblique impact crater indicated by zone of avoidance ray pattern defining Palus Somni (Chart 8). Clementine.

Taruntius (B7, 57 km): Floor-fractured crater with faint composition rays, thus 1-2 b.y. old. Concentric mountainous ring, rilles and dark ashy deposits are visible on floor.

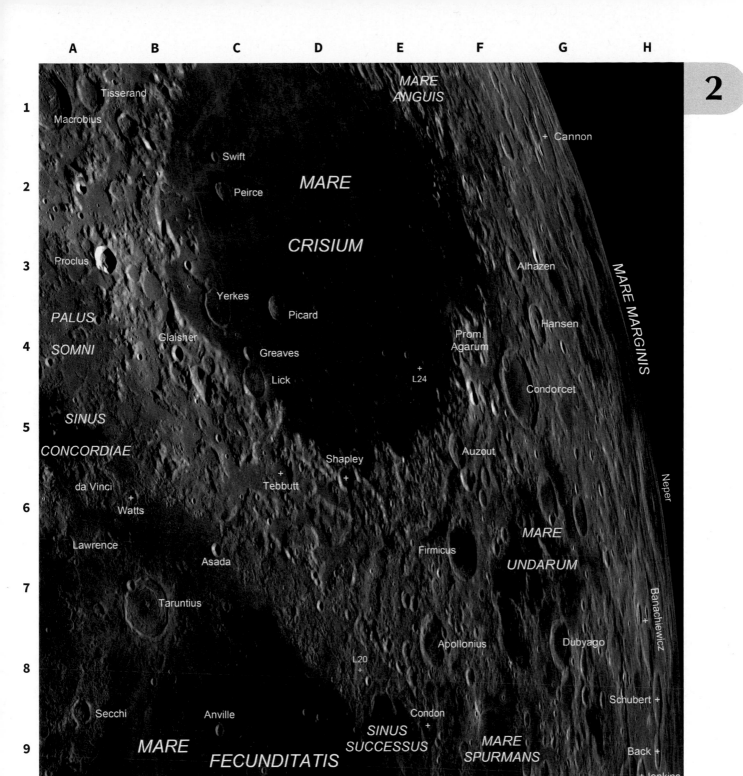

MARE ANGUIS

Tisserand

Macrobius

+ Cannon

Swift

Peirce

MARE

CRISIUM

Proclus

Alhazen

Yerkes

Picard

Hansen

PALUS

Glaisher

Greaves

Prom. Agarum

Condorcet

SOMNI

Lick

+ L24

MARE MARGINIS

SINUS

Shapley

Auzout

CONCORDIAE

+ Tebbutt

+

da Vinci

+ Watts

Neper

MARE

Lawrence

UNDARUM

Asada

Firmicus

Taruntius

Banachiewicz +

Apollonius

Dubyago

L20 +

Schubert +

Secchi

Anville

Condon +

Back +

MARE

SINUS SUCCESSUS

MARE SPURMANS

Nobili + +Jenkins

FECUNDITATIS

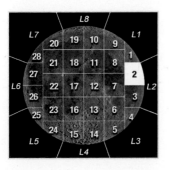

Crisium 2

Lunar 100:

L10 - Crisium (D3)
L12 - Proclus (A3)
L31 - Taruntius (B7)
L100 - Marginis Swirls (G2)

Maria typically fill circular impact basins, but there is little evidence for a basin under Mare Fecunditatis. It is a topographic low, so presumably its basin rim has been reduced to insignificance by erosion. The most dramatic feature here is the Copernicus-like crater Langrenus, and Mare Smythii is right on the limb. The most famous feature is the Messier pair. View the entire Fecunditatis and Smythii basins and their mare ridges on Charts B4 and B2.

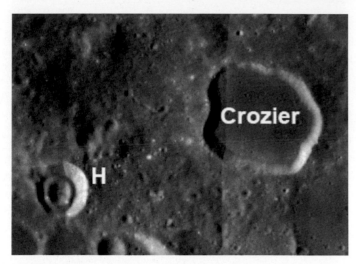

Crozier H (B6, 11 km): Second easiest concentric crater for visual observation (after Hesioidus A; Chart 16). Small, requiring high power to detect inner ring.

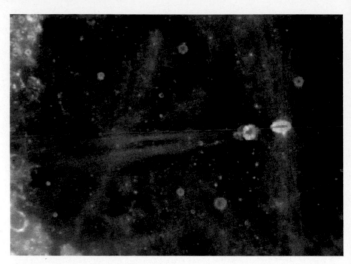

Messier and **A** (B1, 14 & 11 km): The most bizarre crater pair on the Moon, formed by a grazing impact with projectile coming from the east. Elongated Messier was excavated and Messier A made by ricocheted projectile. Two parallel rays to west of Messier A, and very faint butterfly rays to north and south of Messier are visible with high Sun.

Goclenius Rilles (A5, 190 km): Stunning Apollo 8 images made famous these rilles that cross crater floor and rim and then continue to northwest. Are they tension cracks resulting from subsidence of the Mare Fecunditatis basin?

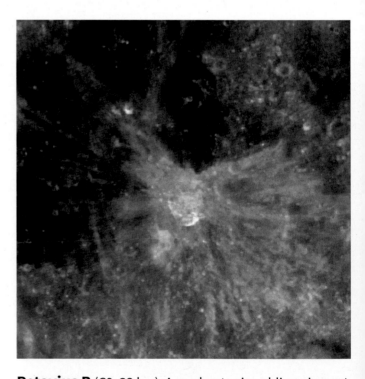

Petavius B (C9, 32 km): A moderate size oblique impact crater with distinct rayless zone of avoidance to north, indicating the direction of the incoming projectile.

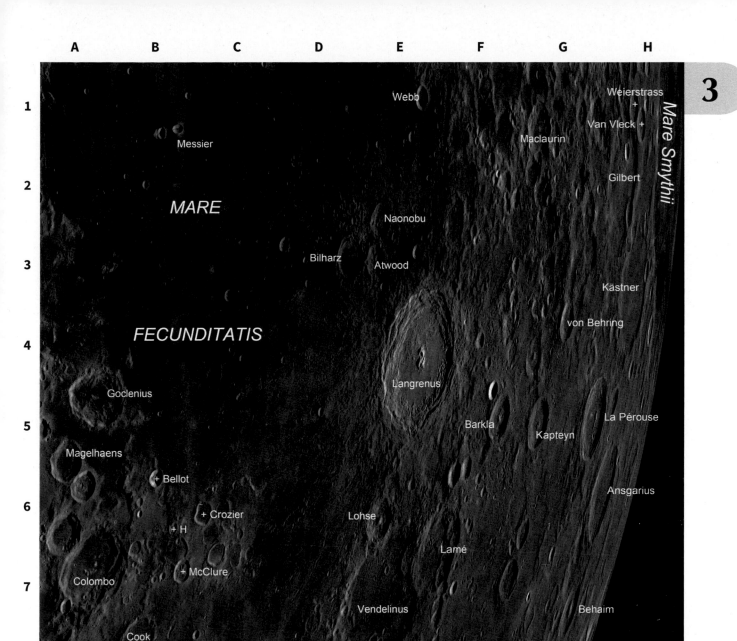

A B C D E F G H

1 — Webb — Weierstrass +

Messier — Van Vleck +

Mare Smythii

Maclaurin

2 — *MARE* — Gilbert

Naonobu

3 — Bilharz — Atwood — Kästner

von Behring

4 — *FECUNDITATIS* — Langrenus

5 — Goclenius — Barkla — Kapteyn — La Pérouse

Magelhaens

6 — + Bellot — Lohse — Ansgarius

+ Crozier

+ H

7 — + McClure — Lamé

Colombo — Vendelinus — Behaim

8 — Cook — + Gibbs

Monge — Holden

Balmer

9 — Petavius B — + Hecataeus

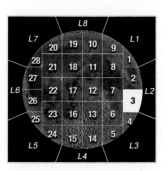

Fecunditatis 3

Lunar 100:

L25 - Messier and Messier A (B1)
L85 - Langrenus rays (D4)

The southeast corner of the Moon is squeezed between the Nectaris and Australe basins and a possible one at Fecunditatis. At one time this area was covered by overlapping deposits of ejecta from these old basins, but now only one piece of evidence is prominent – the Snellius Valley that is radial to Nectaris. The main attractions are two large craters that are quite similar, but unequally visible. Petavius and Humboldt are both floor-fractured craters with floors crossed by rilles and dark volcanic deposits. Unfortunately, the larger and more spectacular crater is near the limb and only occasionally decently visible; but Petavius is a memorable sight itself!

Furnerius (A6, 135 km): Smaller than Petavius, with a shallow floor, low peak and worn walls that demonstrate it is older. A patch of younger mare and a rille enliven the floor.

Humboldt (F3, 200 km): Squeezed by foreshortening into a narrow ellipse, dark lavas cling to NE and SW walls, and three bright peaks parade across the middle of the floor.

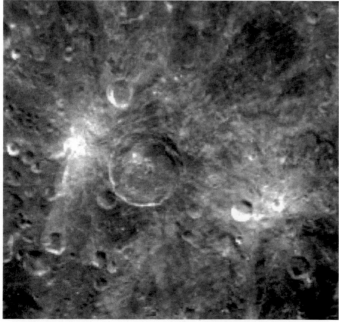

Stevinus A (A4, 10 km) and **Furnerius A** (B5, 10 km): Two very bright small craters look like headlights at full Moon with surprisingly long rays for such small craters. Stevinus crater is between them.

Snellius Valley (A3 - C5, 640 km): Less conspicuous than the Rheita Valley (Chart 5), the Snellius Valley is also radial to the Nectaris Basin, and the crateriform shape of its components show that it is a secondary crater chain from the basin. (Visualization of LRO altimetry data)

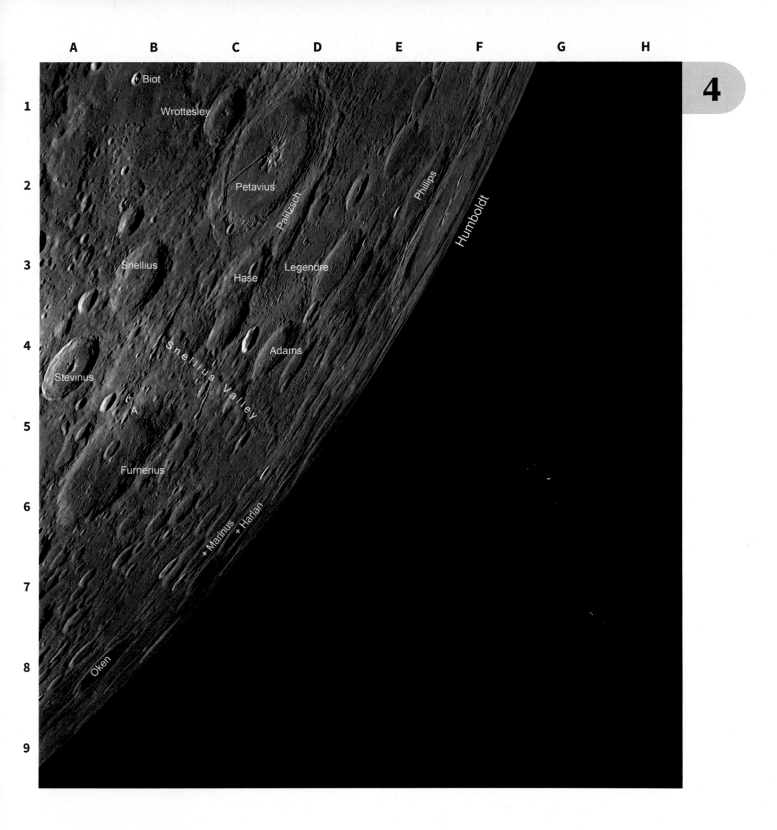

+ Biot

Wrottesley

Petavius

Palitzsch

Phillips

Humboldt

Snellius

Legendre

Hase

Adams

Snellius Valley

Stevinus

+
A

Furnerius

+ Marinus + Harlan

Oken

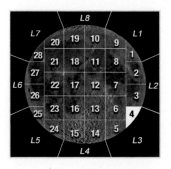

Petavius 4

Lunar 100:

L16 - Petavius (C2)
L87 - Humboldt (F2)

This is a region of highland terrain with an oddly rimless impact basin – Australe – straddling the limb. The multi-generational crater complex of Janssen and its floor's unusual highland rille are near top-center of the chart. The long Rheita Valley is the best nearside example of basin secondary crater chain such as is common on the farside radiating from the Orientale Basin. View the Australe Basin on Chart B3.

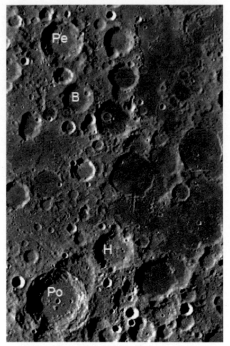

Mare Australe (G6, 612 km): Impact basins typically are giant depressions surrounded by at least partial rim segments. Australe visually isn't a depression and no basin rim is visible, but the circular distribution of mare-floored craters and low spots between craters suggests that this is an ancient impact basin whose original rim subsided because the early Moon's crust was too warm to support massive rims. Pe = Peirescius, B = Brisbane, H = Hanno, Po = Pontecoulant.

Janssen (C3, 201 km): Janssen has at least two oddities – it isn't circular and it has a large rille, which is rare for the highlands. Janssen is elongated and its northern part has a lower rim and a rougher floor than the southern part. Janssen is actually two or more craters. The large curved Janssen Rille cuts Fabricius ejecta and smooth plains on the southern crater floor. Is this material non-mare volcanism or basin ejecta?

Rheita Valley (F1-F4, 509 km) A disintegrating mountain ejected during the formation of the Nectaris Basin dropped 10 or more gigantic masses whose impacts created the Rheita Valley secondary crater chain. The reduction in crater diameter and change in trend limbward of Mallet encourage the unlikely speculation that the two segments are not related.

Fabricius (D3, 79 km): If all craters followed the standard model we'd be bored; Fabricius keeps us guessing. Why does it have two off-center mountain ridges rather than a normal central peak complex? Is the longer, western ridge connected to the wall by a large slump mass?

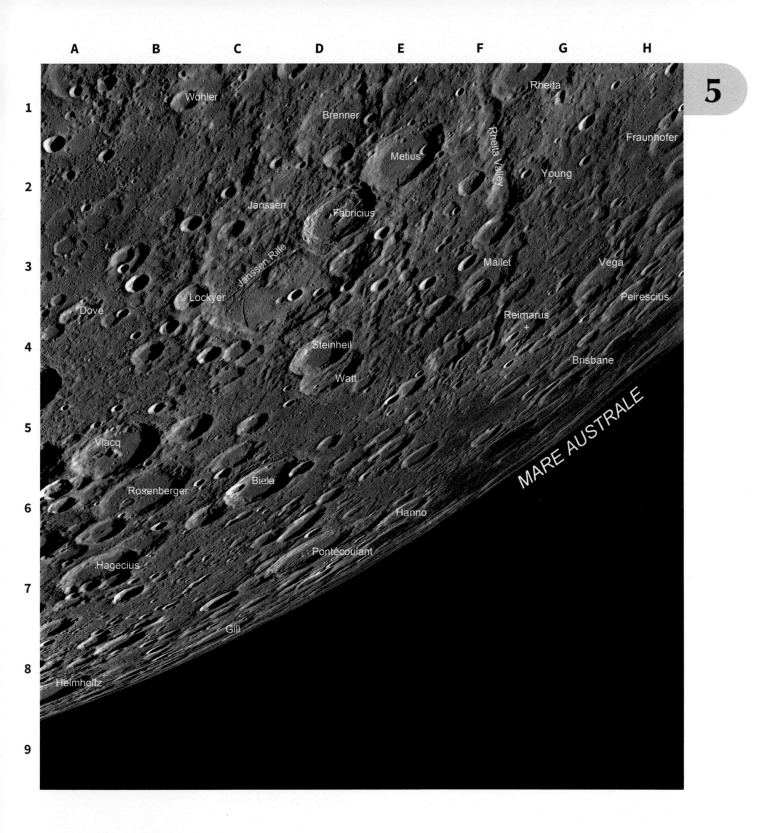

MARE AUSTRALE

Janssen 5

Lunar 100:

L40 - Janssen (C3)
L56 - Australe (G6)
L58 - Rheita Valley (F1-F4)

Impact basins are considered to be wet or dry, depending on how much mare lava they contain. Imbrium is very wet, and Nectaris is quite dry. Only a small amount of mare lava fills the central part of Nectaris, but patches of relatively smooth gray, not dark, material between the Altai ring and the inner rings may be older lavas. View the Nectaris Basin and its mare ridges on Chart B4.

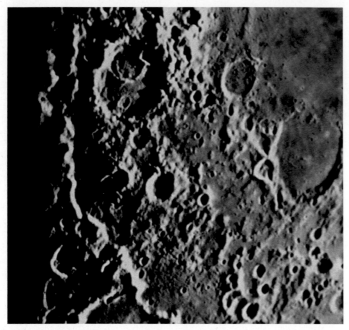

Altai Scarp (A4-C7, 550 km): The Apennines (Chart 18) and the Altai are the two most spectacular basin rims visible from Earth. The Altai Scarp rises 3-4 km above the adjacent terrain and 7 km above the basin center. Careful observers can trace it eastward past Piccolomini to Borda. (CLA)

Bohenberger (G2, 33 km): Perhaps because there is so little mare in Nectaris there is only one floor-fractured crater. The hilly center of Bohenberger was broken by small fractures as it domed upward.

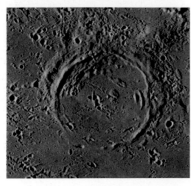

Daguerre (E1, 45 km): An almost ghost crater with faint rim arcs, and dark pyroclastic material visible under high Sun. Is the entire crater volcanic or is there just a vent near an old impact crater?

Fracastorius (E4, 128 km): Originally a large complex crater, now tilted towards the center of the basin. Mare lavas over-topped the northern rim, flooding the floor. After the lavas cooled and solidified, the basin floor subsided more, creating a hairline crack across the crater floor. Nearby Beaumont (C3) had a similar history of being tilted inward, but no crack.

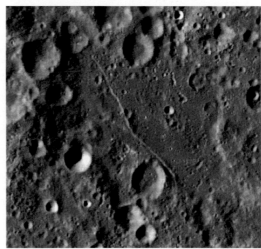

Neander Fault (G8, 77 km): Long overlooked, a rare highland fault cuts basin ejecta filling an ancient unnamed crater.

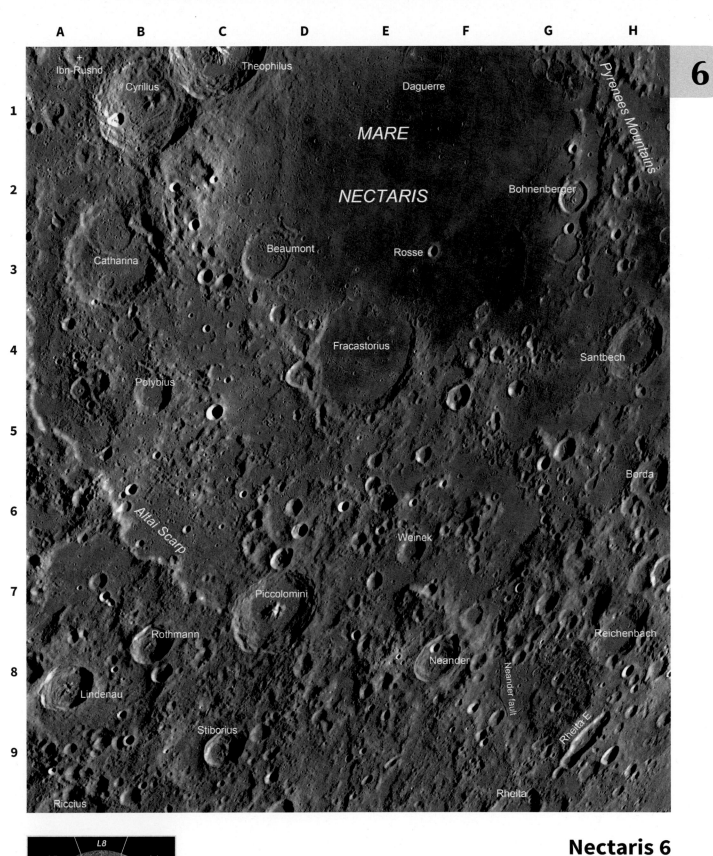

MARE

NECTARIS

A B C D E F G H

1
2
3
4
5
6
7
8
9

Ibn-Rushd
Cyrillus
Theophilus
Daguerre
Pyrenees Mountains
Bohnenberger
Catharina
Beaumont
Rosse
Fracastorius
Santbech
Polybius
Altai Scarp
Borda
Weinek
Piccolomini
Rothmann
Reichenbach
Neander
Neander fault
Lindenau
Rheita E
Stiborius
Rheita
Riccius

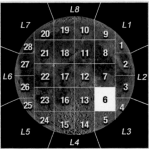

Nectaris 6

Lunar 100:

L7 - Altai Scarp (A4-C7)
L8 - Theophilus, Catharina, and Cyrillus (C1-B3)
L21 - Fracastorius (E4)

Tranquillitatis

Mare Tranquillitatis is famous as the place where humans first walked on the Moon: "Tranquillity Base. The Eagle has landed." In addition to the Apollo 11 site, (and nearby precursor landings of Ranger 8 and Surveyor 5 – see Chart LS1), this mare hosts classic volcanic domes near Arago and Cauchy, and the latter crater is bounded by a rille and a rare lunar fault. At the bottom of Chart 7 is one of the Moon's grand craters, 100 km wide Theophilus and its impact melts. View the entire odd Tranquillitatis Basin and its mare ridges on Chart B5.

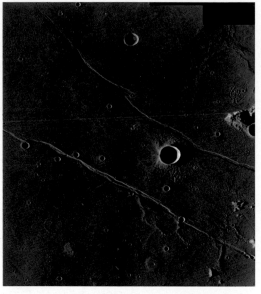

Cauchy Rille, Fault and **Domes** (G1): An area rich in volcanic and tectonic features, all difficult to see well. Cauchy (12 km) is a bowl-shaped simple crater with a curved rille to its north and the famous Cauchy Fault to the south. The fault is ~170 km long and rises ~340 m high. Two pancake-shaped domes lie south of the fault. Low Sun required.

Sabine (right, A5, 30 km) and **Ritter** (Chart 12, 30 km): Inner hilly rings suggested that these were volcanic craters, but they are now recognized as floor-fractured variants of impact craters.

Lamont (B3, 80 km): Unusual named feature composed entirely of mare ridges and visible only with low Sun. The Lamont oval is partially surrounded by a larger and less complete oval. Mare ridges radiate from Lamont, which is interpreted as a two-ring impact basin covered by mare lavas.

Theophilus (C9, 100 km): A grand impact crater, remarkable for having readily observed impact melts. Typical of a large complex crater, there is a flat floor (melt covered), massive central peak complex, and terraced walls that stair-step down 4.4 km to the floor. Flat ponds of impact melt are visible just outside the crater rim to the north, and faint rays radiate across southern Tranquillitatis.

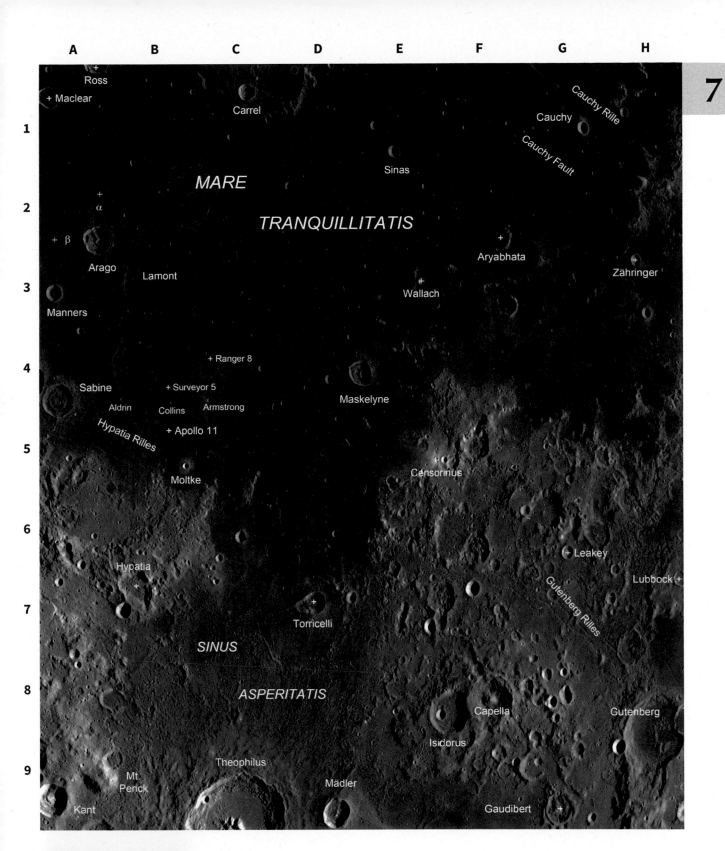

A   **B**   **C**   **D**   **E**   **F**   **G**   **H**

Ross
+ Maclear

Carrel

Cauchy Rille
Cauchy
Cauchy Fault

MARE

Sinas

TRANQUILLITATIS

α
β

Aryabhata

Arago
Lamont

Wallach

Zähringer

Manners

+ Ranger 8

+ Surveyor 5

Sabine
Aldrin   Collins   Armstrong

Maskelyne

Hypatia Rilles   + Apollo 11

Moltke

Censorinus

+ Leakey

Hypatia

Lubbock +

Gutenberg Rilles

Torricelli

SINUS

ASPERITATIS

Capella

Gutenberg

Isidorus

Theophilus

Mt.
Penck

Mädler

Kant

Gaudibert

Tranquillitatis 7

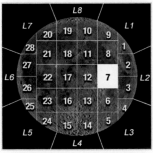

Lunar 100:

L32 - Arago Domes (A2)   L53 - Lamont (B3)
L38 - Sabine & Ritter (A4)   L90 - Apollo 11 (B5)
L48 - Cauchy Fault (G1)

Should be called the Taurus Hills because the topography east of Serenitatis is low and rounded, presumably ejecta from the formation of Serenitatis and more distant basins. The highlights here are Posidonius, the dark edges of the mare, and the Apollo 17 landing site (Chart LS1) – hopefully not the last place ever for human visits to the Moon.

Gardner Megadome (E8, 70km): Wide elevated area, south of Gardner crater (top), possibly a large shield volcano with a caldera on top.

Plinius (B8, 43 km): From its commanding position at the junction of Serenitatis and Tranquillitatis, Plinius seems like a giant fortress. At low Sun its ejecta visibly extend across the maria, showing that the crater is younger. North are the Plinius Rilles, formed as Serenitatis subsided, cracking its edges.

Serpentine Ridge (A3-A7, 365 km): Mare ridges are officially named for geologists (this is the Lister and Smirnov ridges), but the 200+ year old name Serpentine Ridge is delightfully descriptive and takes precedence. The winding ridge is part of 620 km diameter inner ring of the Serenitatis Basin.

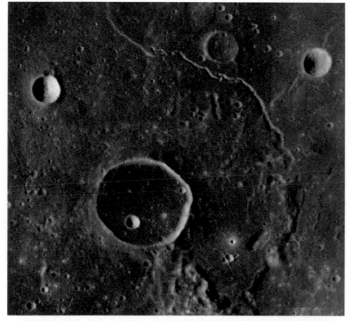

Jansen (D9, 23 km): Jansen crater is nearly filled to its rim with mare lavas compositionally different from its surroundings. The nearby rille suggests this is a volcanic complex.

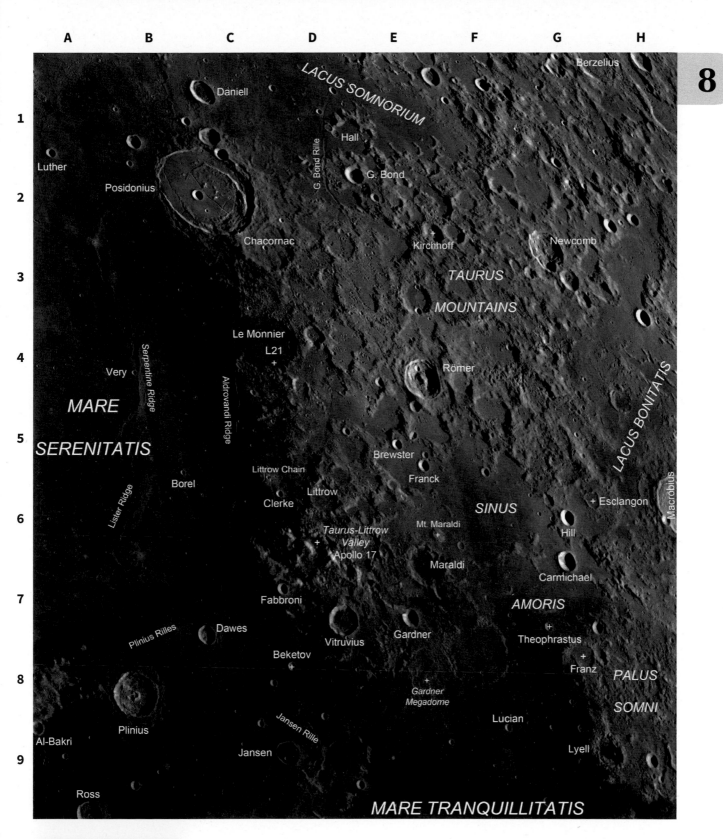

A B C D E F G H

8

Berzelius

1

Luther

Daniell

LACUS SOMNORIUM

Hall

G. Bond Rille

Posidonius

2

G. Bond

Chacornac

Kirchhoff

Newcomb

TAURUS

3

MOUNTAINS

Le Monnier

L21

4

Very

Serpentine Ridge

Römer

LACUS BONITATIS

MARE

Aldrovandi Ridge

5

SERENITATIS

Brewster

Franck

Macrobius

Littrow Chain

Lister Ridge

Borel

Littrow

SINUS

Esclangon

6

Clerke

Taurus-Littrow
Valley
Apollo 17

Mt. Maraldi

Hill

Maraldi

Carmichael

Fabbroni

AMORIS

7

Plinius Rilles

Dawes

Vitruvius

Gardner

Theophrastus

Beketov

Franz

PALUS

8

Gardner
Megadome

SOMNI

Plinius

Jansen Rille

Lucian

Al-Bakri

Lyell

9

Jansen

Ross

MARE TRANQUILLITATIS

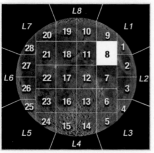

Taurus Mountains 8

Lunar 100:

L18 - Serenitatis Dark Edges (A8, C5)
L20 - Posidonius (C2)
L33 - Serpentine Ridge (A3-A7)

Two Greek heroes are at the center of this region, with the massive Humboldtianum Basin anchoring the limb. The eastern end of Mare Frigoris intrudes on the left, with little ponds of maria at Lacus Temporis and the darker Lacus Spei. The largest crater is the Plato-like Endymion, and the larger, but shared with Chart 10, Lacus Mortis is a crater with a lava name. See Chart B1 for a full view of the Humboldtianum Basin.

Atlas (D7, 87 km) and **Hercules** (C7, 71 km). A favorite lunar pair because they are close in location and size, but have different interiors. The larger Atlas is a floor-fractured crater with rilles and two dark areas of pyroclastic deposits from volcanic dark halo craters. Hercules has been modified by volcanism in a more prosaic way – lavas cover its floor. Atlas overlaps and subdues a smaller crater to the north.

Franklin (F9, 56 km) and **Cepheus** (F8, 39 km): A smaller version of Atlas and Hercules, with the larger Franklin being floor-fractured and overlapping a smaller crater to the south.

Endymion (D5, 122 km): A large, smooth-floored crater – the Archimedes of the northeast limb - is a pointer for the main Humboldtium basin ring immediately towards the limb.

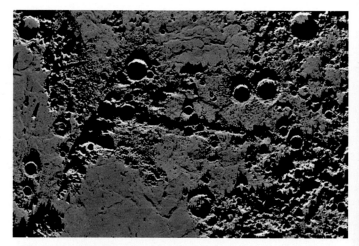

Shannen Ridge (A8-D8, 630 km): Maurice Collins noticed a ridge radial to the center of the Imbrium Basin and stretching from the edge of a mountain block near Eudoxus (Chart 10), along the southern edge of Lacus Mortis, and continuing from Mason to Williams. This ridge, unofficially named for Maurice's daughter, aligns to the west with the abrupt offset between the Alpes and Caucasus mountains; the ridge appears to mark a major, but poorly understood, Imbrium fracture. Image visualized from LRO altimetry.

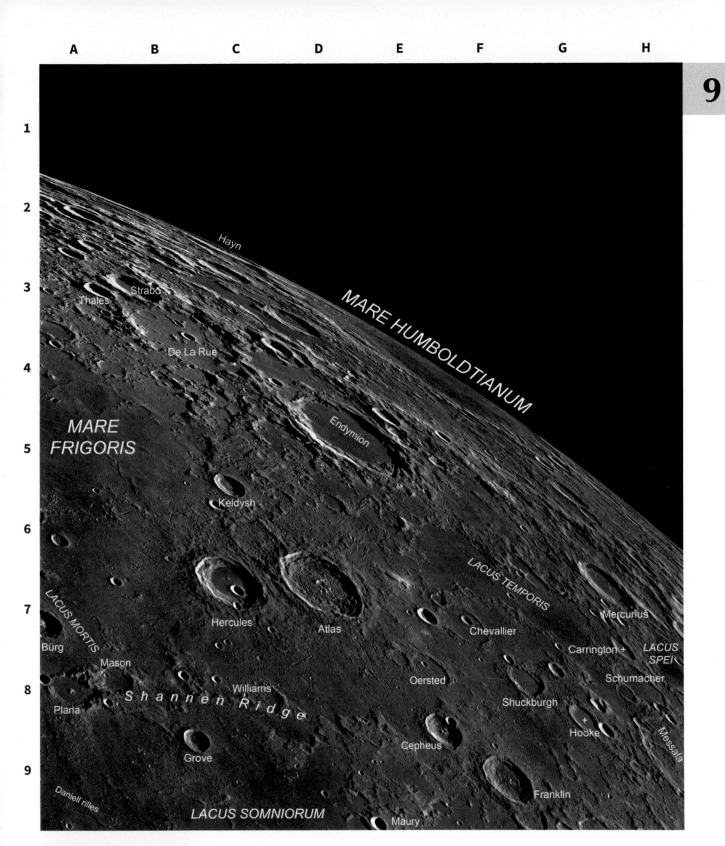

A B C D E F G H

1

2

Hayn

3 Strabo
Thales

De La Rue

MARE HUMBOLDTIANUM

4

MARE
FRIGORIS

Endymion

5

Keldysh

6

LACUS TEMPORIS

LACUS MORTIS

7 Mercurius
Bürg
Hercules Atlas
Chevallier
Carrington +
LACUS
SPEI

Mason

Schumacher

8 Williams
S h a n n e n   R i d g e
Oersted
Shuckburgh
Plana +
Hooke
Messala

Cepheus
Grove

9

Daniell rilles
Franklin

LACUS SOMNIORUM
Maury

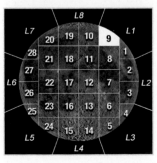

Humboldtianum 9

Lunar 100:

L70 - Humboldtianum (F4)
L72 - Atlas & Hercules (C7-D7)

The eastern side of the lunar prime meridian, approaching the North Pole, is a region with many favorite landforms. North of Mare Frigoris is mostly battered craters, pummeled and infilled by ejecta from the formation of the Imbrium Basin. The bottom half of the sheet includes two divergent rim segments of the basin, the Alpes and the Caucasus mountains, as well as the most spectacular linear valley on the Moon, the Alpine Valley. Nearby are two large young impact craters Aristoteles and Eudoxus, and the western half of Lacus Mortis – the half with rilles and a fault.

Alpes Mountains (A7, 335 km): The Alpes are a peculiar piece of the Imbrium Basin's rim. The side facing Mare Imbrium is very mountainous, creating a front similar to the Apennines, and behind the front are much smaller hills that are progressively smaller away from the front.

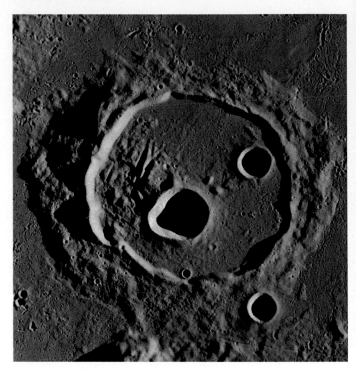

Cassini (B9, 56 km): Strangely appealing crater shallowed with lava and 2 craters on the floor an odd, squashed looking mountain lies between west rim and largest crater on floor.

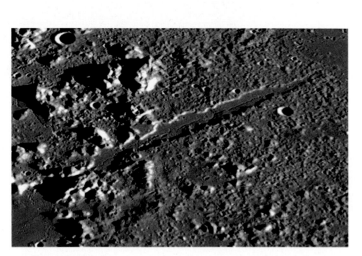

Alpine Valley (A6, 155 km): The Alpine Valley is a long split in the Alpes Mountains. It points back toward the center of the Imbrium Basin and appears to be where the crust outside the basin was fractured and split apart during adjustment to the impact. The Valley's floor is lava-filled, erupted from a challenging to observe, narrow sinuous rille in its middle.

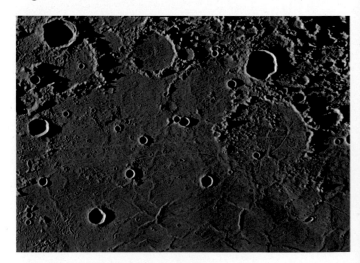

Sheepshanks-Gärtner Rille (G4, 200 km): Two names, but one delicate rille connects two craters in this otherwise rille-free northern zone. The Gärtner part is easier to see.

A   B   C   D   E   F   G   H

1

Main
Challis
De Sitter
Scoresby
Euctemon
Baillaud
Meton
Petermann
Barrow
Cusanus

2

Neison
W. Bond
Moigno
Arnold
Schwabe
Kane
C. Mayer
Democritus

3

Sheepshanks
Gärtner
Sheepshanks Rille
Gärtner Rille
Archytas

4

Protagoras
Galle

MARE FRIGORIS

5

6

Alpine Valley
Baily
Mitchell
Egede
Aristoteles

ALPES

7

LACUS
MORTIS
Bürg
Bürg Rille

8

Eudoxus

Alexander

9

Cassini
CAUCASUS MOUNTAINS
Calippus
Theaetetus

Aristoteles 10

Lunar 100:

L19 - Alpine Valley (A6)
L26 - Frigoris (B4)
L34 - Lacus Mortis Fault (H8)
L76 - W. Bond (A2)

L8
L7   20   19   **10**   9   L1
28   21   18   11   8   1
27   22   17   12   7   2
L6   26   23   16   13   6   L2
25   24   15   14   5   4
L5   L4   L3   3

Mare Serenitatis is a large, nearly featureless expanse of lava with only the 16 km wide crater Bessel significantly breaking the surface. The basin's rim is largely missing, with the Haemus Mountains being the most continuous remnant. Immediately south is a low area raked by Imbrium ejecta, creating striated ridges and ruined crater rims, interspersed with small mare ponds. Two major segments of the Imbrium Basin rim – the Apennine and Caucasus mountains cut across the northwest corner. View the entire Serenitatis Basin on Chart B5.

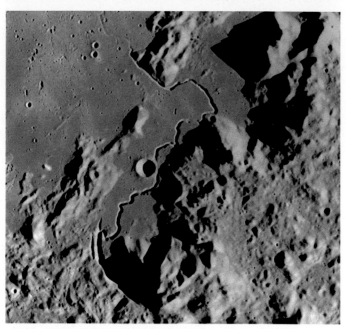

Hadley Rille (B4, 116 km): The most dramatic Apollo landing site required dropping down over the peaks of the Apennine Mountains and landing next to a sinuous rille. The astronauts were 3.3 billion years too late to witness molten lava flowing down the rille as an active lava channel. Apollo 15 image.

Sulpicius Gallus Rilles (E6, 80 km) These shore-hugging short rilles in older Serenitatis lavas were created by the subsidence of the basin center loaded by the mass of newer lavas. Menelaus Rilles (H7) have the same origin, as do the Plinius and Littrow Rilles (Chart 8) to the east.

Linné (E4, 2 km): A tiny crater with a huge history, Linné was reported in 1866 to have disappeared. It is a young crater but not that young! Until photos could accurately document the surface, there was no way to determine if reported changes were real. Linné didn't change, observers erred. LRO NAC image.

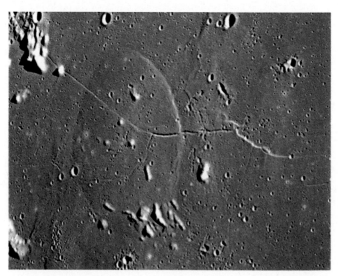

Valentine Dome (D3, 30 km): A nearly flat plateau cut by a just observable fault/rille continuing in from the mare. Visible only under very low illumination. Apollo 15 image.

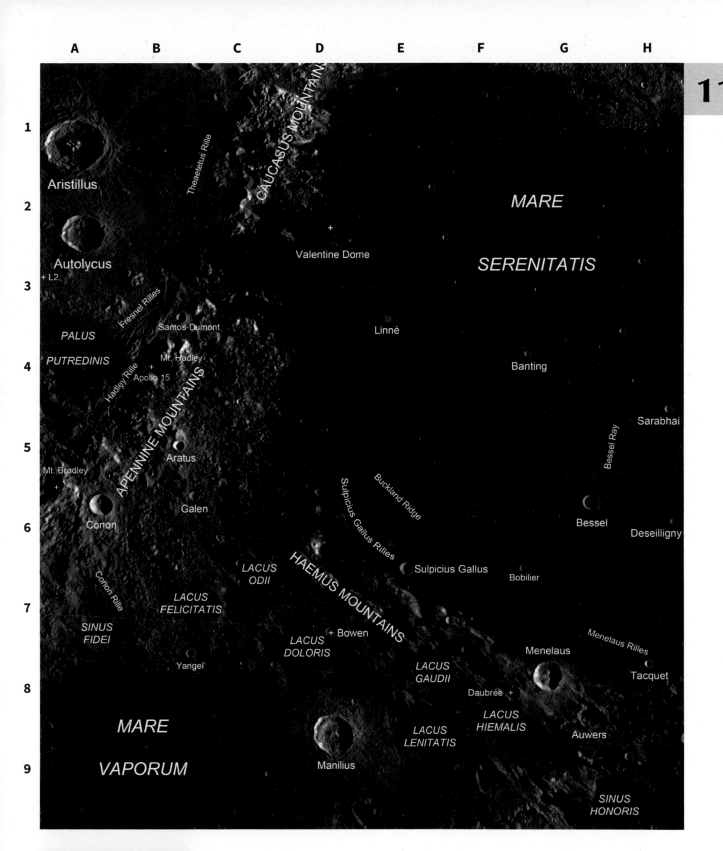

A **B** **C** **D** **E** **F** **G** **H**

1

Aristillus

CAUCASUS MOUNTAINS

Theaetetus Rille

2

Autolycus

+ L2

MARE

SERENITATIS

Valentine Dome

3

Fresnel Rilles

PALUS

PUTREDINIS

Santos-Dumont

Linné

Banting

Mt. Hadley

Hadley Rille

+ Apollo 15

APENNINE MOUNTAINS

4

Sarabhai

Bessel Ray

5

Aratus

Mt. Bradley

Galen

Buckland Ridge

Bessel

Deseilligny

6

Conon

Sulpicius Gallus Rilles

Sulpicius Gallus

Bobilier

Conon Rille

LACUS ODII

HAEMUS MOUNTAINS

7

LACUS FELICITATIS

SINUS FIDEI

Menelaus Rilles

+ Bowen

LACUS DOLORIS

Menelaus

Tacquet

Yangel'

LACUS GAUDII

8

Daubrée +

Auwers

LACUS HIEMALIS

MARE

LACUS LENITATIS

9

VAPORUM

Manilius

SINUS HONORIS

Serenitatis 11

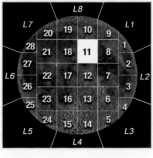

Lunar 100:

L41 - Bessel Ray (H5)
L66 - Hadley Rille (B4)
L71 - Sulpicius Gallus
       Dark Mantle (E6)

L82 - Linné (E4)
L89 - Valentine Dome (D2)
L99 - Ina (B7)

Highlands cover most of the Moon, and this chart includes the northern extension of the classic southern highlands of the nearside. The area is dominated by large, old, ruined craters, with a sprinkling of brighter, smaller, newer ones. Across the northern edge are remnants of craters striated by Imbrium ejecta. Between the pasty ejecta and ancient craters are light-hued smooth plains (another part of Imbrium's ejecta) cut by some of the Moon's most famous rilles. In the southeast corner the rubbly Descartes Plains was the landing site for Apollo 16.

Descartes Mountains (G9, D = 38 km) : Apollo 16 was a mission to the highlands but the knobby hills partially covering Descartes (Chart 13) are very unusual even for the highlands. Most sampled rocks are fragmental ejecta, not volcanic.

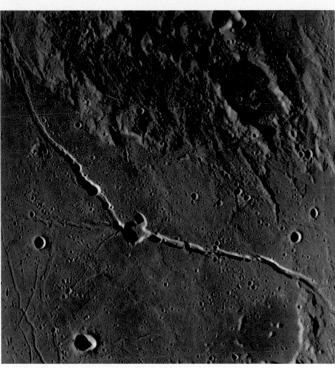

Hyginus Rille (C2, 204 km): Hyginus is a rimless depression at the junction of two large crater-pit rilles – it is probably the largest volcanic crater on the Moon.

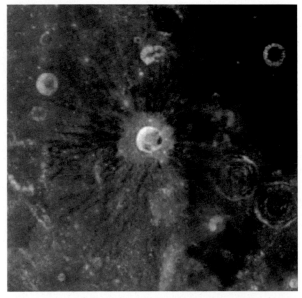

Dark-Rayed Dionysius (H4, 140 km): Under a high Sun this crater is surrounded by bright collar and dark rays. Dionysius formed on the basin ejecta - mare boundary; dark rays are mare material excavated from under the bright basin ejecta. (Clementine)

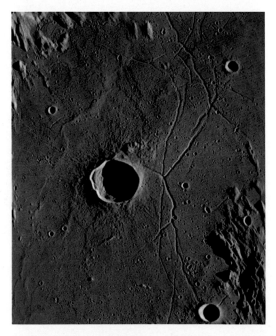

Triesnecker Rilles (B3, 200 km): K-shaped nexus of rilles just east of the complex crater Triesnecker. Network of older rilles to north just connects to the Hyginus Rille.

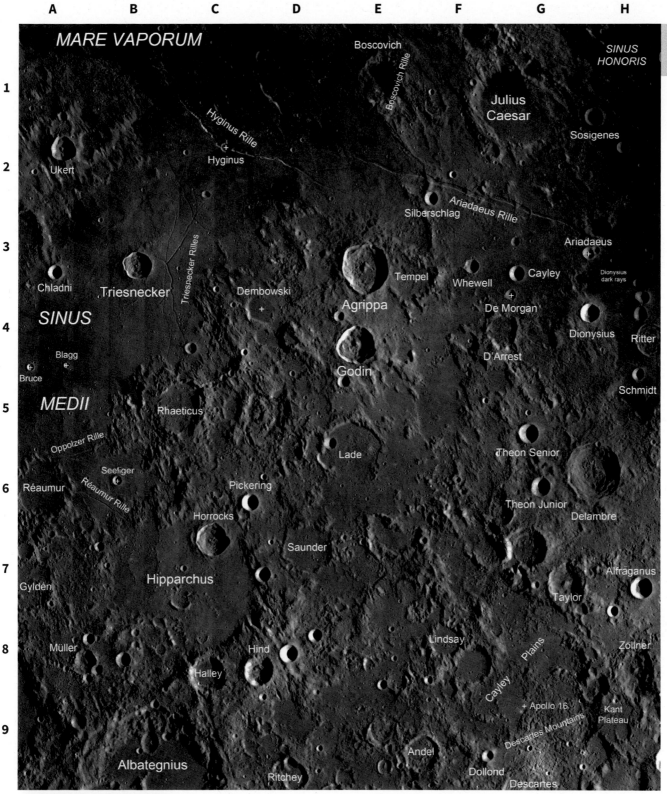

MARE VAPORUM

Boscovich

SINUS HONORIS

Julius Caesar

Sosigenes

Ukert

Hyginus Rille

+ Hyginus

Boscovich Rille

Ariadaeus Rille

Silberschlag

Ariadaeus +

Chladni

Triesnecker

Triesnecker Rilles

Dembowski +

Tempel

Whewell

Cayley

Dionysius dark rays

Agrippa

De Morgan +

SINUS

Blagg

Bruce

Godin

Dionysius

Ritter

MEDII

Rhaeticus

D'Arrest

Schmidt

Oppolzer Rille

Lade

Theon Senior

Seeliger

Réaumur

Réaumur Rille

Pickering

Theon Junior

Delambre

Horrocks

Saunder

Gyldén

Hipparchus

Alfraganus

Taylor

Müller

Hind

Lindsay

Zöllner

Halley

Cayley Plains

+ Apollo 16

Kant Plateau

Albategnius

Andel

Descartes Mountains

Ritchey

Dollond

Descartes

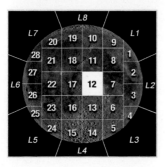

Hipparchus 12

Lunar 100:

L24 - Hyginus Rilles (C2)

L28 - Hipparchus (B7)

L29 - Ariadaeus (H3)

L35 - Triesnecker (B3)

L38 - Sabine & Ritter (H4)

L50 - Cayley Plains (G8)

L64 - Descartes (G9)

L93 - Dionysius Dark Rays (H4)

This is the central nearside highlands with many degraded craters and light-hued smooth plains – possibly fluidized basin ejecta - filling crater floors and spaces between craters. No single feature dominates but the craters Werner and Abulfeda at opposite sides bound the region. The concentric crater Pontanus E is near the center of a very degraded possible ancient impact basin (Werner-Airy) with a diameter about 500 km. It is easy to get lost in this region of similar looking craters!

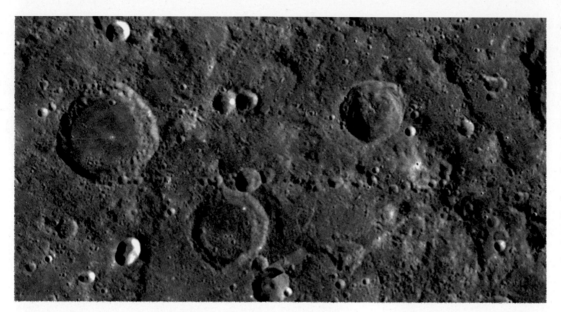

Abulfeda Crater Chain (F1-H3, 210 km): About two-dozen small craters define a long line tangent to the rim of Abulfeda. Most crater chains are secondary craters radial to their primary, but there is no primary for these, which may have resulted from the impact of a disintegrating comet. Previous interpretations as a volcanic fracture with small eruptions or collapses along it are inconsistent with the lack of evidence of volcanism in this area.

Pontanus E Concentric Crater (E6, 13 km): Virtually all concentric craters occur on or near maria except this one. E lies within the putative Werner-Airy basin, consistent both with an interpretation that concentric crater inner rims are volcanic, and that the basin may be real and contain ancient non-mare volcanism.

Werner (B7, 71 km): Young, but rayless, complex crater with small off-center "central" peaks and smaller hills. A very small but bright impact pit on the north wall shines brightly at full Moon.

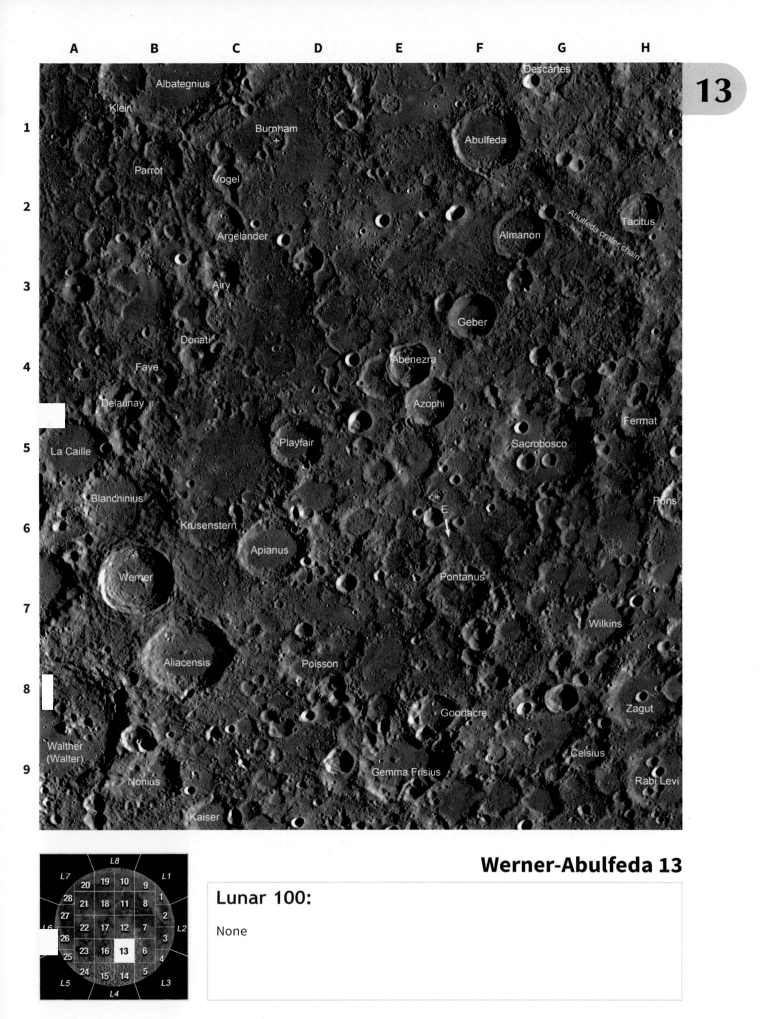

Werner-Abulfeda 13

Lunar 100:

None

Classic highlands with old, flat-floored craters. This area has a surplus of 20 to 30 km wide flat-floored craters, suggesting that many are secondary craters from the formation of the Orientale and perhaps Imbrium basins. There are no dramatic young craters here, but the concatenation of overlapping craters near the South Pole is impressive. Slightly rubbled smooth areas between craters hints that fluidized basin ejecta or non-mare volcanism filled and smoothed over earlier heavily cratered crust.

Boussingault (G9, 128 km): A favorite near-limb crater within a crater caused by a subsequent random impact making an eccentric inner crater.

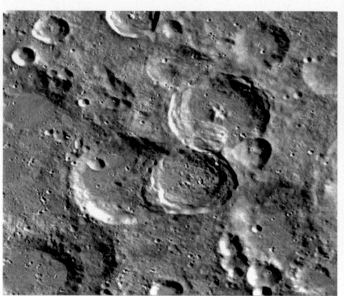

Miller (A1, 61 km), **Nasireddin** (A2, 50 km) and **Huggins** (Chart 15, H2, 66): Three relatively large overlapping craters, with Nasireddin causing a huge landslide into Miller and a smaller one into Huggins. (M3 image)

Heraclitus (B4, 85 km): A strange elongated feature that may be multiple craters, or, with its central ridge, an older version of Schiller, formed by a low oblique impact.

Schomberger (C9, 88 km): A Copernicus look-alike hidden near the limb at 77°S. Too close to the pole to know if it has rays, but probably not.

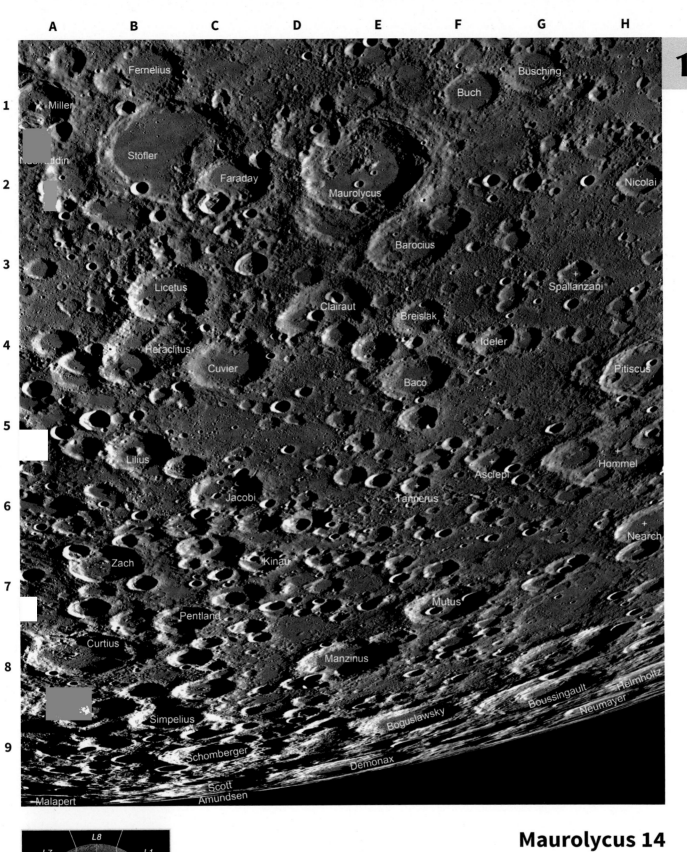

Maurolycus 14

Lunar 100:

L45 - Maurolycus (E2)
L55 - Baco (F4)
L96 - Leibnitz Mountains (A9)

This is the most dramatic part of the nearside highlands because of the spectacular craters Tycho, Clavius and Moretus, shadow-filled deep craters near the South Pole, and the blocky mountains seen in projection at the limb. Site of the explorative crash-landing of the LCROSS probe in 2010 which confirmed that water ice does exist in permanently shadowed polar craters.

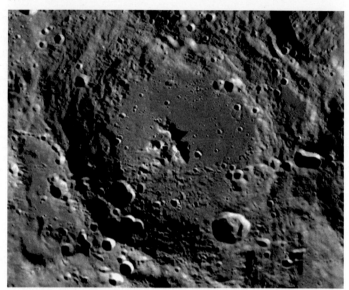

Drygalski (D9, 162 km): Older Copernicus twin visible nearly in profile on limb when seen at all.

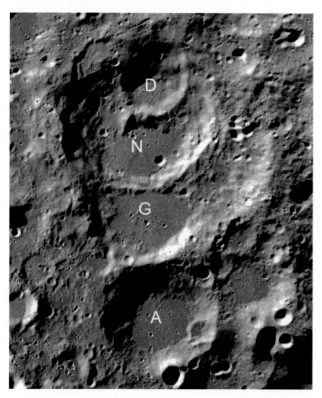

Newton (G9, 75 km): Hyper-famous scientist with barely visible battered limb crater. The floor of Newton D is 7 km below a mountain on its rim – one of the greatest elevation differences on the Moon. – and lower than the floors of Newton (N) and Newton G and A.

Moretus (H8, 114 km): A foreshortened Tycho clone with similar terraces and peak but the lack of a dark collar and rays marks it as older.

Tycho (E2, 86 km): A prototype of a complex crater. Tycho has grand terraces that step 4.7 km down from the serrated rim crest to a flattish floor, surfaced with impact melt and littered with hills and a massive central peak complex. Strong roughness around the crater is thick ejecta veneered with melt that makes a dark halo at full Moon. The longest rays on the nearside radiate from Tycho, but they are nearly absent on the west, indicating formation by an oblique impact.

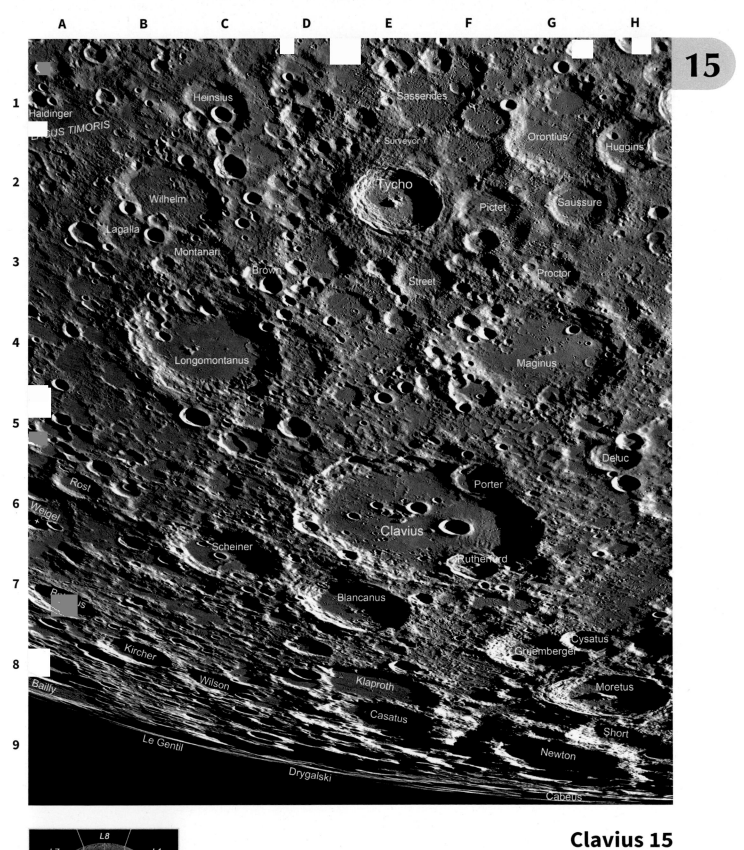

A B C D E F G H

1 Haidinger
LACUS TIMORIS
Heinsius
Sassenides
+ Surveyor 7
Orontius
Huggins

2 Wilhelm
Tycho
Pictet
Saussure
Lagalla

3 Montanari
Brown
Street
Proctor

4 Longomontanus
Maginus

5 Deluc

6 Rost
Weigel +
Scheiner
Clavius
Porter
Rutherfurd

7 Blancanus
Cysatus

8 Kircher
Gruemberger
Moretus
Wilson
Klaproth
Bailly
Casatus
Short

9 Le Gentil
Newton
Drygalski
Cabeus

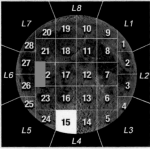

Clavius 15

Lunar 100:

L6 - Tycho (E2)
L9 - Clavius (E6)
L94 - Drygalski (D9)

The lavas of Mare Nubium fill an ancient impact basin (Chart B7) for which little evidence remains. The center of the basin subsided, tilting Pitatus inward, and on the east, faulting through the middle of an old crater (Ancient Thebit), submerging its western rim and creating the Straight Wall in its middle. On the west of Nubium a line of partial crater rims mark a high ridge of older features surrounded by maria. At upper right is one of the most controversial craters on the Moon, Alphonsus, which almost certainly did not erupt in 1959, despite Russian observations.

Alpetragius (G2, 40 km): Normal rim, but rounded peak looks like a giant egg in a bird's nest.

Hesiodus A (B7, 14 km): A large concentric crater with an inner donut ring that is hard, but satisfying, to glimpse.

Alphonsus (G1, 111 km): Ranger 9 crashed into the northeast floor of Alphonsus providing close-ups of the famous volcanic dark halo craters. A meridianal ridge of Imbrium ejecta drapes the floor, and despite claims, the ancient anorthosite central mountain did not erupt in 1959.

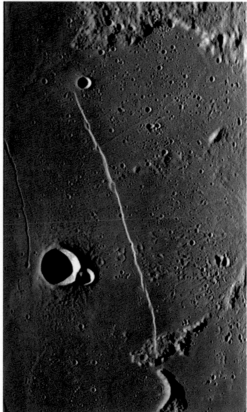

Straight Wall (E4, 116 km): This 400 m high scarp is the Moon's most visible fault that is not a basin rim. The east side is high, casting a thin dark shadow at sunrise. Just west is the delicate Birt Rille and dome. At low Sun observe the mare ridge rim of Ancient Thebit, the informal name for the 200 km wide crater containing the Wall.

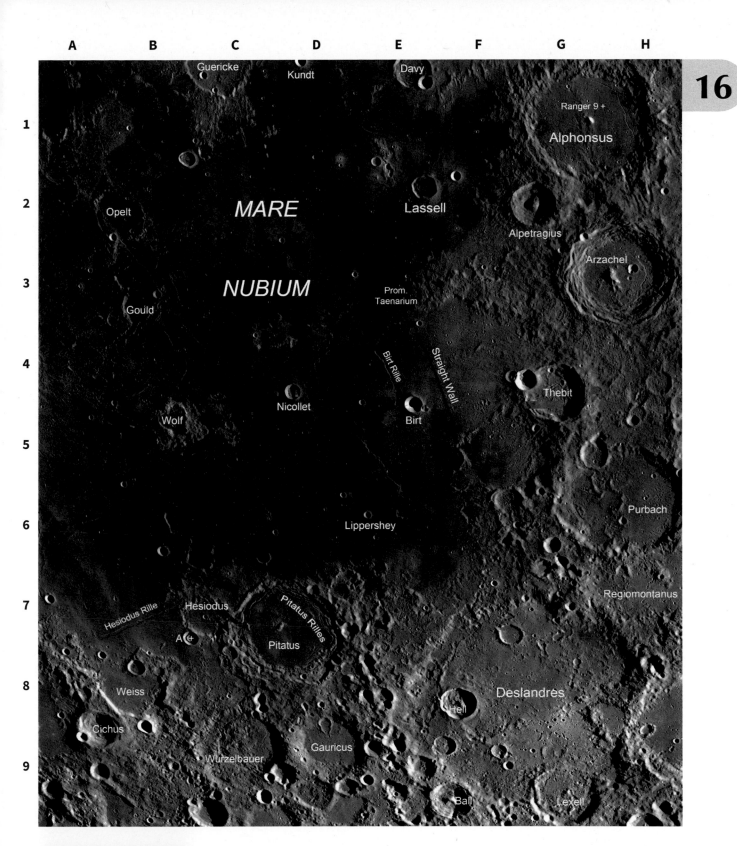

Nubium 16

Lunar 100:

L15 - Straight Wall (E4)

L46 - Regiomontanus central peak (H7)

L47 - Alphonsus (G1)

L81 - Hesiodus A (B7)

L84 - Pitatus (C7)

The stars of this sheet are at opposite corners. Copernicus is described on Chart 22, and Ptolemaeus and nearby Herschel are at bottom right. The rest of the sheet is dominated by thin lavas filling low spots between mounds of ejecta from the Imbrium Basin impact. Apollo 14 landed north of Fra Mauro crater to collect ejecta samples that determined the formation age of the basin – 3.85 billion years ago.

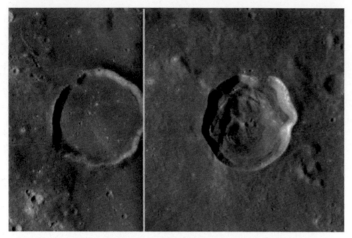

Gambart (B5, 25 km) and **Mösting** (F5, 25 km): Two craters of the same size with equal-looking rim crests, but totally different interiors. Mösting is a typical transition complex crater with a small rebound peak and slumps – slides of debris down the walls – that are not yet terraces. Did Gambart look like Mösting and later was flooded with lava? Or did Gambart never look like that and thus have a different origin? There are other Gambart-like craters in this area (Tobias Mayer, Kunowsky) so is what is special about it?

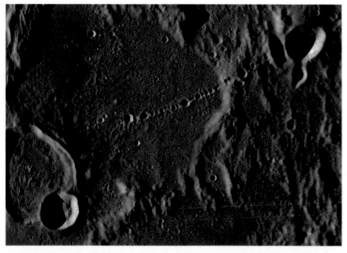

Davy Crater Chain (F9, 52 km): Long considered a certain volcanic feature; presumably eruptions along a fracture formed more than a dozen small craters. But other than its alignment there is no strong evidence that the chain is volcanic. After Comet Shoemaker-Levy sheared into numerous fragments and crashed into Jupiter it was proposed that the Davy Chain formed the same way.

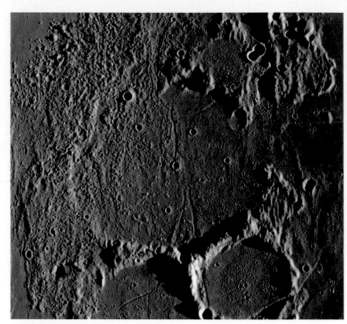

Fra Mauro (B8, 97 km): Copernicus-size old crater filled with rubbly ejecta from formation of Imbrium Basin – somewhere under ejecta are terraces and central peaks.

Copernicus H (A1, 4 km): One of multiple small dark halo craters near Copernicus. Not volcanic craters, but normal post-Copernicus impacts that bring up dark underlying mare rock as dark ejecta on top of bright Copernican rays. (M3 image)

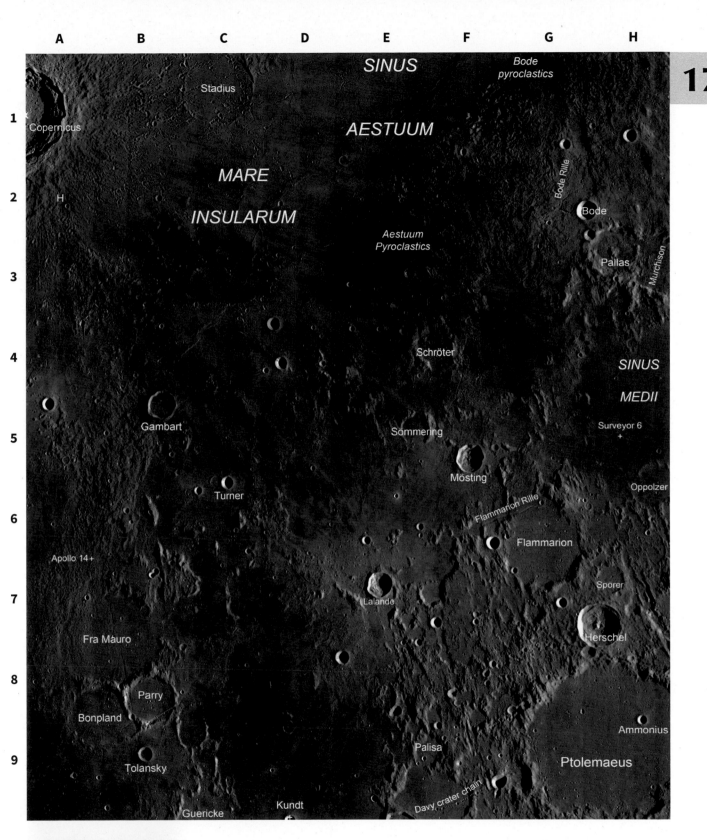

A B C D E F G H

SINUS

Bode pyroclastics

1 Copernicus
x
Stadius

AESTUUM

2 H

MARE

INSULARUM

Bode Rille

Bode

Aestuum
Pyroclastics

Pallas

Murchison

3

4 Schröter

SINUS

MEDII

Gambart

5 Sömmering

Surveyor 6
+

Turner

Mösting

Oppolzer

6 Flammarion Rille

Flammarion

Apollo 14 +

7 Lalande

Spörer

Fra Mauro

Herschel

8 Parry

Bonpland

Ammonius

9 Palisa

Tolansky

Ptolemaeus

Guericke

Kundt
+

Davy crater chain

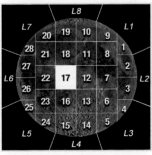

Ptolemaeus 17

Lunar 100:

L5 - Copernicus (A1)   L74 - Copernicus H (A2)
L17 - Fra Mauro (B7)   L75 - Ptolemaeus (H9)
L51 - Davy Crater Chain (F9) L92 - Gyldén Valley (H7)
L61 - Mösting A (F6)

A classic lunar area for determining stratigraphy or age sequences. The age progression goes from ancient (Apennine basin rim), to younger (Archimedes formed on Imbrium Basin floor) to younger (Mare Imbrium lavas) to youngest (secondary craters/rays from Copernicus).

Archimedes (G3, 81 km): Originally a Copernicus look-alike but its external ejecta was covered by late Mare Imbrium lavas, and central peaks and wall terraces were buried by Imbrium age lavas that rose up fractures under the crater.

Apennine Bench (G5, 280 x 350 km): Geologists' name for terrain from Archimedes Mountains to the Apennine Mountains – pieces of the original floor of the Imbrium Basin high enough to not be covered by later Imbrium lavas.

Bode Pyroclastics (G9, 85 km): One of the Moon's largest volcanic ash deposits, dark at full Moon. Ash erupted from V-shaped vent and perhaps the rille. Extends into Chart 17.

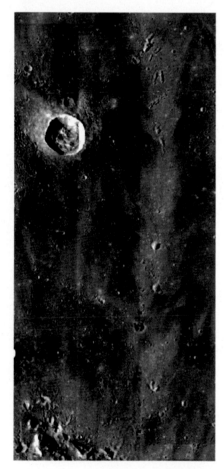

Copernicus Ray and Secondary Craters (A7): Rays and small but detectable secondary craters whose ejecta contribute to ray brightness cross Mare Imbrium. This ray from Copernicus is to the east of Pythias.

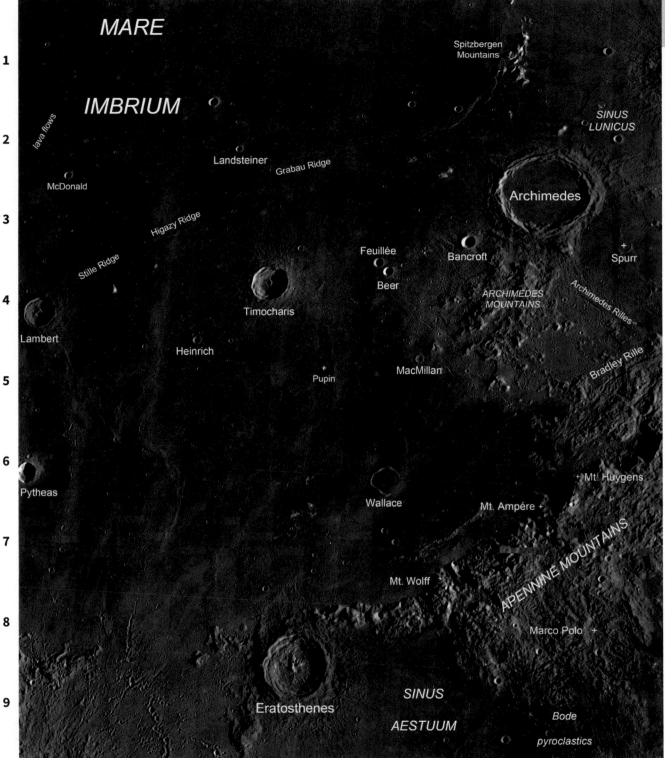

MARE

IMBRIUM

lava flows

McDonald

Landsteiner

Grabau Ridge

Higazy Ridge

Stille Ridge

Lambert

Timocharis

Heinrich

+ Pupin

Feuillée

Beer

Bancroft

ARCHIMEDES MOUNTAINS

Archimedes

SINUS LUNICUS

+ Spurr

Archimedes Rilles

Bradley Rille

MacMillan

Pytheas

Wallace

+ Mt. Huygens

Mt. Ampére +

Mt. Wolff +

APENNINE MOUNTAINS

Marco Polo +

Eratosthenes

SINUS AESTUUM

Bode pyroclastics

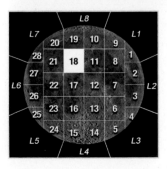

		L8		
L7	20	19	10	9	L1
28	21	18	11	8	
27	22	17	12	7	L2
L6	26	23	16	13	6	
25	24	15	14	5	
L5		L4		L3

Archimedes 18

Lunar 100:

L4 - Apennine Mtn. (G7)

L27 - Archimedes (G3)

L69 - Copernicus Ray & Secondary Craters (A6-B9)

L78 - Lambert R (A5)

L79 - Sinus Aestuum (E9)

From northern Imbrium past Plato and Mare Frigoris to the North Pole is one of the most scenic areas of the Moon. The grand star is Plato, distinctive with its bright and shadow-casting rim, and smooth dark floor. The hilly and pitted terrain from east of Plato to west of Sinus Iridum (Chart 20) is composed of layers of ejecta; Imbrium's is oldest and is covered by Iridum's and then Plato's. The terrain beyond Frigoris is largely veneered by hilly or smooth Imbrium ejecta that has inundated all but the post-basin craters such as Philolaus and Anaxagoras. Near the pole are mostly shallow, medium size craters (e.g. the pair Challis and Main) lacking the drama of the opposite pole.

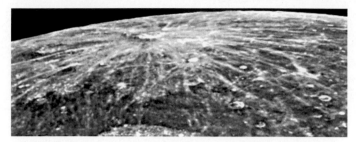

Anaxagoras (G1, 51 km): Rayed crater with paucity of rays to the west, suggesting an oblique impact. CLA.

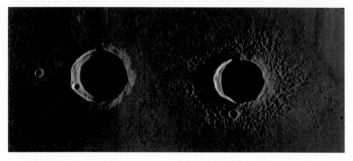

Helicon (left; A9, 24 km) and **Le Verrier** (B9, 21 km): The two largest craters in northern Imbrium look similar, but low Sun observing reveals that Le Verrier's ejecta covers the surrounding mare, but Helicon's is covered by it and thus is older than the latest mare lavas and its neighbor.

Plato (F6, 101 km): One of the most viewed craters of the Moon, because it is easy to find, has had multiple reports (all doubtful) of mists and fogs, and is beautiful. A slump of the western wall separated a huge triangular block from the rim but it did not collapse onto the crater floor. The lava-filled floor looks like a giants' skating rink; sharp-eyed observers may see the largest of the pits on its floor.

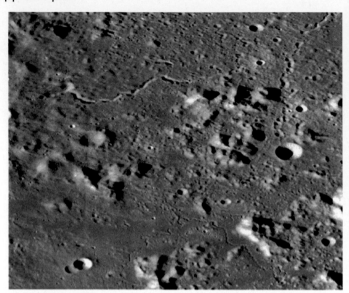

Plato Rilles (H5, 180 km): A variety of sinuous rilles cut Imbrium rubble east of Plato; only the northernmost being readily visible.

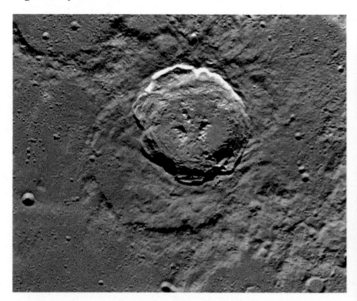

Philolaus (D2, 70 km): Relatively fresh, but rayless complex crater overlapping and filling a 90 km wide earlier ring. Impact melt fills eastern side of floor, with two large separated peaks and smaller ones, rather than a single central massif.

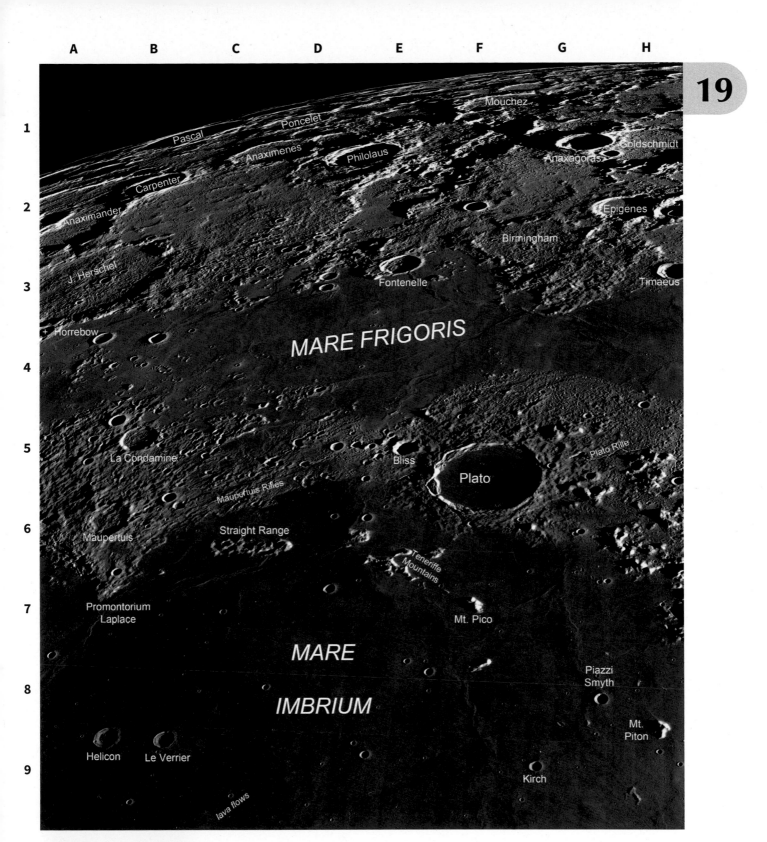

Mouchez

Pascal

Poncelet

Anaximenes

Philolaus

Goldschmidt

Anaxagoras

Carpenter

Epigenes

Anaximander

Birmingham

J. Herschel

Fontenelle

Timaeus

+ Horrebow

MARE FRIGORIS

La Condamine

Bliss

Plato Rille

Plato

Maupertuis Rilles

Straight Range

Maupertuis

Teneriffe
Mountains

Promontorium
Laplace

Mt. Pico

MARE

Piazzi
Smyth

IMBRIUM

Mt.
Piton

Helicon

Le Verrier

Kirch

lava flows

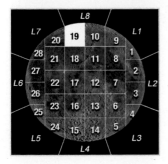

Plato 19

Lunar 100:

L23 - Pico (F7)
L26 - Mare Frigoris (A4-H4)
L83 - Plato Pits (F6)

This northwest curved corner of the Moon includes most of a very conspicuous and spellbinding feature, and a much harder to see but important volcanic dome complex. The intersection of Mare Frigoris with the northernmost part of Oceanus Procellarum somehow received a name, Sinus Roris, but is not of significance. Pythagoras is a bold crater near the limb, and nearby are other Greeks, Xenophanes and Cleostratus, and other mathematicians, Boole and Cremona. Going south the limb becomes more electric with the craters Volta and Galvani.

Mairan T (C8, 7 x 9 km): A small, steep-sided volcanic cone with an irregular crater on top. Difficult and rarely detected.

Pythagoras (G3, 145 km): A large Copernicus-like crater seen nearly in profile (when librations are poor) with massive central peaks and classic terraces. Impact melt ponds to the southeast suggest an oblique impact from the opposite direction.

Sinus Iridum (G8, 249 km): The beautiful Bay of Rainbows looks like a safe harbor with a few lazy waves (mare ridges) rolling in from the mighty Sea of Showers (Mare Imbrium). Iridum is a large impact crater whose southeast side tilted or faulted toward the center of Imbrium and was covered by later mare lavas. The Heraclides headland (bottom left) tapers down in height, suggesting a tilting, but the opposite Promontorium Laplace (Chart 19) is massive, giving evidence for faulting that down-dropped Iridum's missing rim.

Rümker (A9, 73 km): An unusually wide volcanic complex made of a dozen or more overlapping small domes, Rümker is a unique structure on the Moon. It is low and difficult to see without a low Sun. Early mappers thought it a ruined crater.

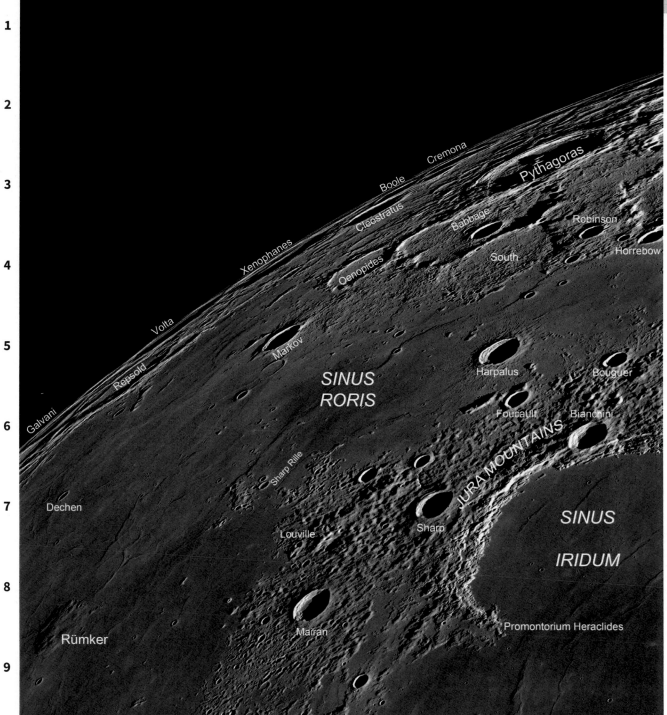

Iridum 20

Lunar 100:

L21 - Sinus Iridum (G8)

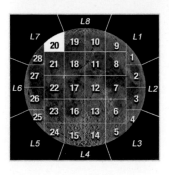

Although no large structural feature dominates this western part of the Mare Imbrium it is still a region of many treasures. The Carpathian Mountains, the extension of the Apennine rim of the Imbrium Basin, is the only obvious evidence that a large basin underlies the area. The Gruithuisen Domes at the north extreme of the image lie along the circular extension of the Apennine - Carpathian rim, and the magmas that created the domes may have risen along deep basin fractures. Isolated mountains such as Vinogradov, Delisle and La Hire are also probably connected to Imbrium Basin rings but the exact relations are not always certain. The Harbinger Mountains are an area of uplift, like the Aristarchus Plateau (Chart 28), and the source of many rilles. Two other volcanic features deserve mention – the long, young lava flows near La Hire, and the large, flat domes near Tobias Mayer.

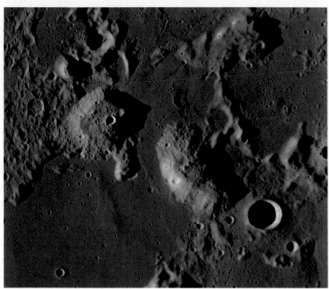

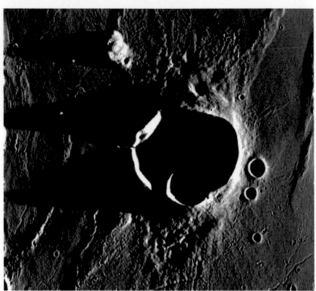

Gruithuisen Domes (D1, 20 & 27 km): Most lunar domes are low and gentle sloped; these two are steep-sided and tall and thus are probably made of more viscous magma.

Krieger (A3, 23 km): A small, lava-flooded crater that casts rabbit ears shadows due to a gap in the western rim where a small rille exits.

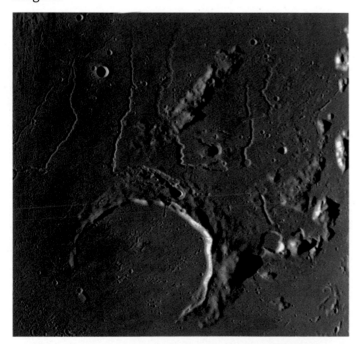

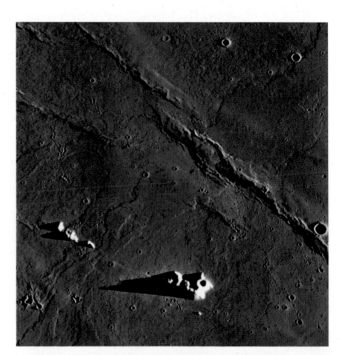

Harbinger Mountains (B4, 93 km) & **Prinz Rilles**: This cluster of hills appears to be uplifted highland rock, like the Aristarchus Plateau. Crustal fracturing permitted magma to reach the surface, forming the five large sinuous Prinz Rilles. Thinner rilles in this family occur to the west.

La Hire Lava Flows (G4, ~400 km): Maria were created by millions of lava flows but individual ones are rare. The easiest to see are three or four that start north of Euler, travel north-east past Mt. La Hire, and cross the Zirkel Ridge, with the flow fronts encroaching into Chart 18.

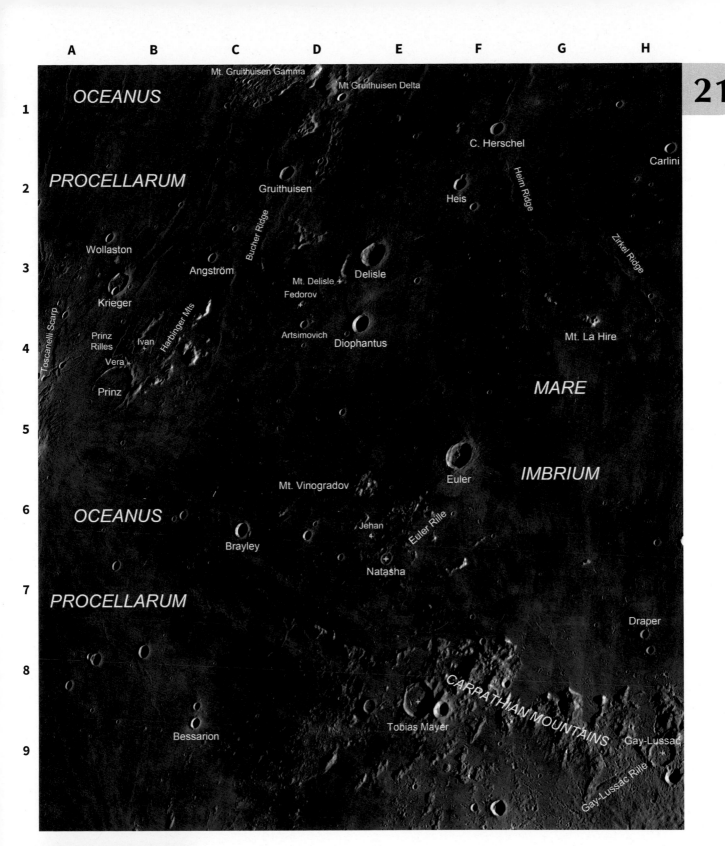

A B C D E F G H

OCEANUS

Mt. Gruithuisen Gamma
Mt Gruithuisen Delta

1

PROCELLARUM

C. Herschel

Carlini

2
Gruithuisen
Heis
Helm Ridge

Zirkel Ridge

Wollaston
Bucher Ridge

3
Angström
Mt. Delisle
Delisle
Fedorov

Krieger

Toscanelli Scarp

Prinz
Rilles
Ivan
Vera
Harbinger Mts
Artsimovich
Diophantus
Mt. La Hire

4

Prinz

MARE

5

IMBRIUM

Euler

Mt. Vinogradov

6
OCEANUS
Jehan
Euler Rille
Brayley
Natasha

7
PROCELLARUM

Draper

8

CARPATHIAN MOUNTAINS

Bessarion
Tobias Mayer
Gay-Lussac

9

Gay-Lussac Rille

Western Imbrium 21

Lunar 100:

L86 - Prinz Rilles (A4)
L98 - Imbrium lavas (G4-H3)

L8
L7   20 19 10   9   L1
     28   1
      21 18 11   8
L7   L2
27   2
L6   22 17 12   7
     26   3
     25 23 16 13   6 4
L5   24 15 14   5
         L4   L3

Half of the most famous crater on the Moon fills a corner of this sheet and its rays and secondary craters occur over at least half the sheet. Nearby Kepler is a miniature Copernicus, and Encke is a shallow version of Kepler. Some of the Moon's most famous domes are near Hortensius and Milichius, and landing sites of four spacecraft are at bottom right, including the victoriously named Mare Cognitum, the Sea That Has Become Known, honoring Ranger 7, the first American lunar probe that worked.

Copernicus (H1, 96 km): This incomparable complex crater has magnificent terraced walls that drop 3.8 km down from the rim crest to the floor. The northwest quadrant of the floor is veneered with impact melt. Scattered central mountains rise about 900 m and smaller hills litter the floor. Looking further beyond the rim are barely visible chains of secondary craters, and at full Moon, bright material blankets the surface a distance about half way to Kepler, with rays extending 800 km.

Hortensius Domes (E2): A classic field of 6 flat-topped domes north of 14 km wide Hortensius. Most of these small volcanoes still have their summit pits. West of here is the Milichius Pi dome, 50% wider than these six.

Encke (B3, 28 km): Polygonal rimmed, floor-fractured and rayless crater. Only 0.7 km deep, compared to 2.7 km depth for the nearly same diameter Kepler (right) that is not floor-fractured.

Kepler (B2, 29 km): Small complex craters like Kepler have masses of slumped wall materials and small clumps of central peaks. A broad bright nimbus and rays surround the crater.

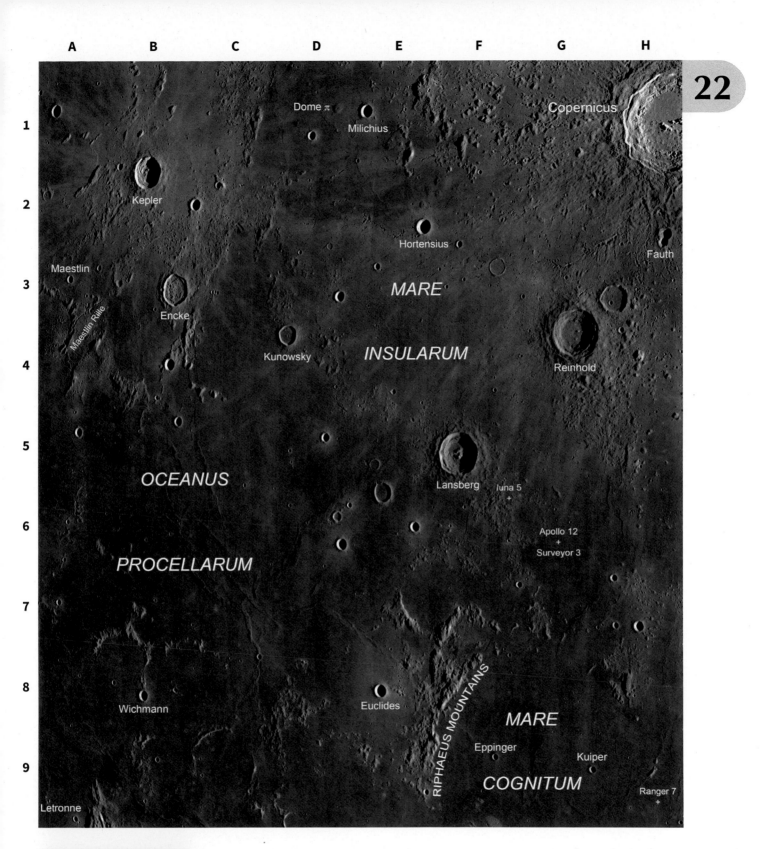

A B C D E F G H

1

Dome π
Milichius
Copernicus

2
Kepler
Hortensius
Fauth

Maestlin

3
MARE
Encke

Maestlin Rille

Kunowsky
4
INSULARUM
Reinhold

5
OCEANUS
Lansberg
luna 5
+

6
Apollo 12
+
Surveyor 3

PROCELLARUM

7

8
Wichmann
Euclides
MARE

RIPHAEUS MOUNTAINS
Eppinger
Kuiper

9
COGNITUM
Ranger 7
+

Letronne

Copernicus 22

Lunar 100:

L5 - Copernicus (H1)
L65 - Hortensius Domes (E2)

Humorum is one of the most instructive basins on the Moon even though most of its ring structure is only weakly visible. The subsidence of the basin as mare filled it tilted Gassendi, Hippalus and Doppelmayer inward, allowing their interiors to be flooded, and also broke the bounding crust with the massive Hippalus RIlles and narrow Doppelmayer Rilles. Palus Epidemiarum and Lacus Excellentiae fill annular low spots in the moat between two Humorum basin rings. Bullialdus and Gassendi are grand craters, with the former nearly pristine and the latter heavily modified. See an overhead view of the Humorum Basin on Chart B7.

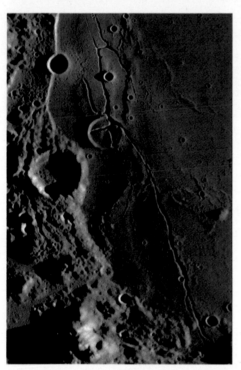

Doppelmayer Rilles (A6-B7, 131 km long): A narrow rille was the source for surrounding dark pyroclastics in southwest Mare Humorum. The southern part of the rille is normal, but the northern portion repeatedly splits into different branches.

"The Helmet" (D2, 60 x 45 km): Light patch of older material made of rare non-mare volcanics, like the Gruithuisen Domes (Sheet 21).

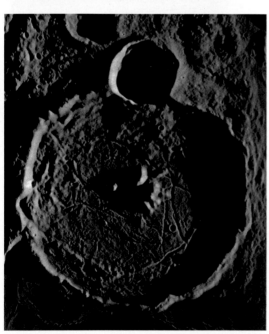

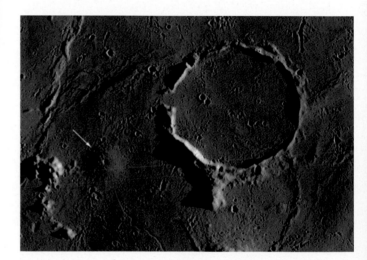

Gassendi (B3, 110 km): Larger than Copernicus and originally similar to it, but now a floor-fractured crater with an uplifted floor crossed with narrow rilles. A tilt toward basin center is revealed by a high rim to north and very low rim at south.

Kies Pi (H6, 14 km): A classic hemispheric dome with summit pit; big and relatively easy to see when the terminator is near. Smaller domes are on the floor of Capuanus (H9).

Norman

Herigonius

Herigonius Rille

Darney

Lubiniezky

Gassendi Rilles

Gassendi

Agatharchides

Bullialdus

MARE

Loewy

König

HUMORUM

Hippalus

Kies

Liebig Scarp

Prom.
Kelvin

Doppelmayer Rilles

Hippalus Rilles

π

Kelvin
Scarp

Palmieri Rilles

Puiseux

Campanus

Mercator Scarp

Palmieri

Doppelmayer

PALUS

Mercator

Lee

Dunthorne

Vitello

Marth

EPIDEMIARUM

Ramsden Rilles

Lepaute

Ramsden

Capuanus

Elger

LACUS
EXCELLENTIAE

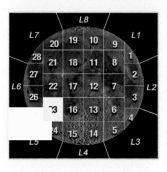

Humorum 23

Lunar 100:

L13 - Gassendi (B3)
L54 - Hippalus Rilles (F6)
L60 - Kies Dome (H6)

Large craters and smooth plains dominate this southwest corner of the Moon. Bailly is a large crater near the limb that appears to be a degraded two-ring basin. Schickard is a landmark in this region for it is large and, like a zebra, has a bright stripe bordered by dark ones. Between Bailly and Schickard is the Schiller-Zucchius Basin (see Chart B8), the largest structure here, but one that was completely overlooked until the 1960s. And Wargentin, one of the most famous peculiar craters, is in the middle of all these other features.

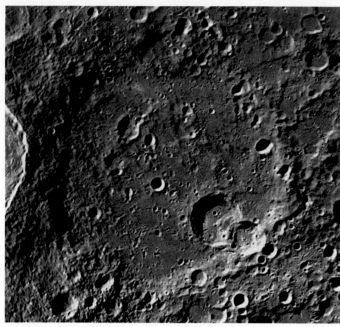

Bailly (H8, 301 km): A broad crater squashed into ellipticity by foreshortening. The walls of Bailly are preserved but the floor with a hint of an inner ring – making it a basin - is broken and highlighted by one large crater at the eastern end. Beyond, when librations are good, is the great Hausen (Chart L5).

Schiller (G5): A weirdly elongated crater, 179 km long but only about 70 km wide. The elongated central ridge in the northern part suggests that Schiller formed by a very low angle impact, a larger version of Messier (Chart 3).

Hainzel (F2, 71 km): Two or three overlapping craters – accident or simultaneous impact? Hainzel is the older crater; half covered by the younger two.

Wargentin (C4, 87 km): The crater famous for being filled to the rim with lava and having a Y-shaped – bird's foot - mare ridge. Lava from the crater cascaded over the northwestern rim, creating the smooth surface west of Schickard.

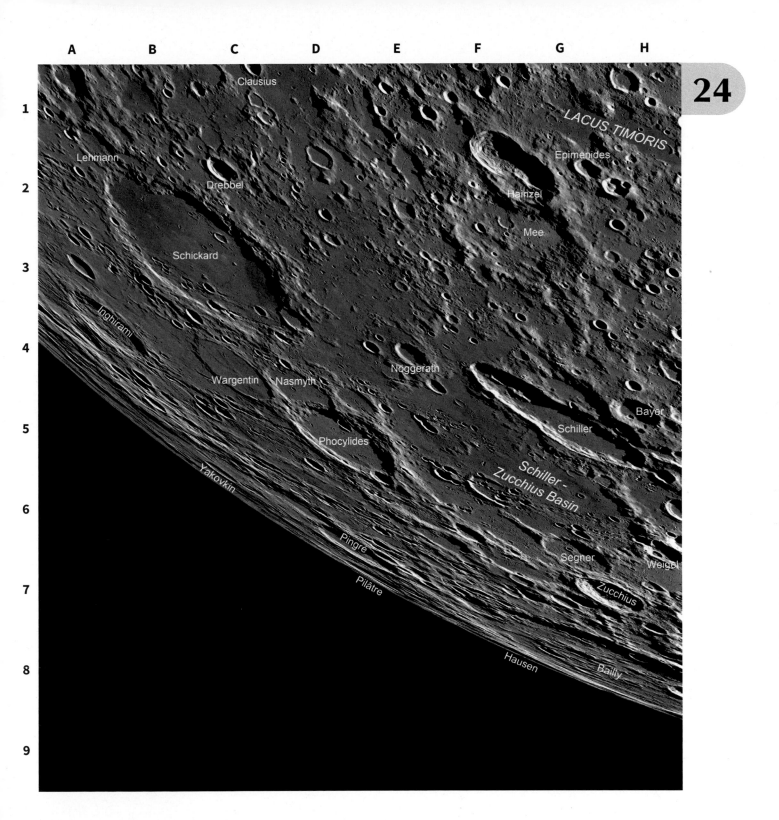

LACUS TIMORIS

Clausius

Lehmann

Drebbel

Epimenides

Hainzel

Mee

Schickard

Inghirami

Noggerath

Wargentin   Nasmyth

Bayer

Schiller

Phocylides

Schiller - Zucchius Basin

Yakovkin

Pingré

Segner

Weigel

Pilâtre

Zucchius

Hausen   Bailly

Schickard 24

Lunar 100:

L30 - Schiller (G5)

L37 - Bailly (H8)

L39 - Schickard (B3)

L43 - Wargentin (C4)

L59 - Schiller Zucchius
Basin (G6)

This corner of the Moon has few well-known craters for a simple reason. It is covered with the most damaging layer of ejecta from the Orientale Basin visible on the nearside. Most of the smaller craters are younger than Orientale and look pristine, but many larger ones, such as Lamarck, Lagrange, Piazzi, and Inghirami are so heavily draped with ejecta that flowed across the surface that they are hardly recognizable as craters. The Cordillera Scarp, visible by the shadow it casts, separates the surrounding terrain from the interior of the Orientale Basin (Charts L6 and B8).

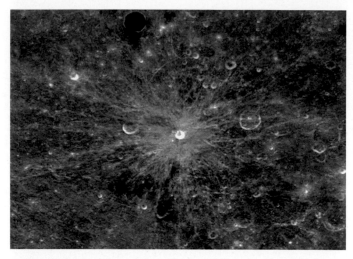

Bygrius A (E1, 18 km): On the eastern rim of ejecta-covered Bygrius is a simple crater with a brilliant ray system that dominates this region under a high Sun. Clementine.

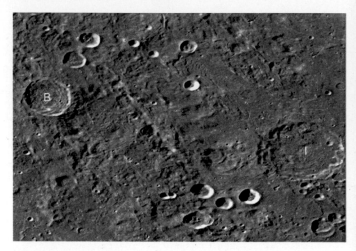

Inghirami Valley (H8, 145 km): Rarely noticed from Earth, between Baade (B) and Inghirami (I) is a massive, degraded crater chain - a broad trough of basin secondary craters filled with massive waves of frozen rock – chain and fill are all Orientale ejecta.

de Gasparis Rilles (H2, 31 km): Within de Gasparis crater is the junction of the continuation of the rilles at Palmieri (Chart 23) with northern rilles probably originally part of Mersenius Rilles (Chart 26).

Unnamed Mare Patch South of Vieta (G5, 89 km): Small patch of dark mare lava is one of a half dozen in this area, as well as the dark ends of Schickard (Chart 24). Is this a mare region only visible where Orientale ejecta are thin? There is no evidence for a buried basin.

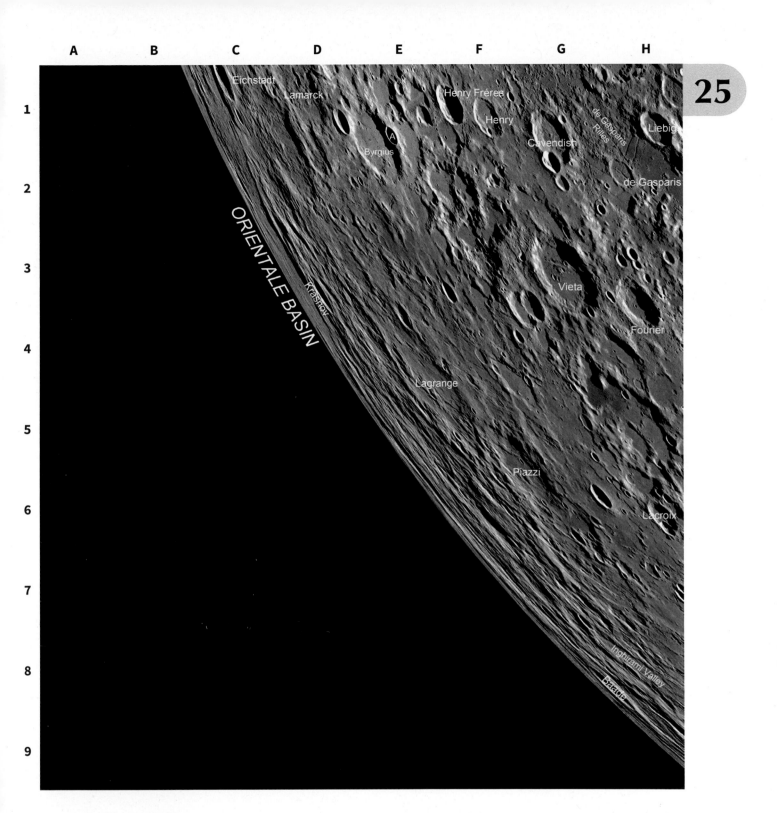

ORIENTALE BASIN

Krashov

Eichstadt
Lamarck
Henry Fréres
Henry
Cavendish
de Gasparis Rilles
Liebig
de Gasparis
Byrgius
A
Vieta
Fourier
Lagrange
Piazzi
Lacroix
Inghirami Valley
Baade

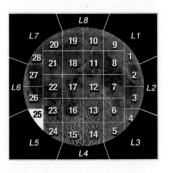

Orientale South 25

Lunar 100:

L91 - de Gasparis Rilles (H2)
L97 - Inghirami Valley (H8)

This is one of the most feature-rich areas of the Moon, being at the intersection of southern Procellarum and Orientale ejecta. Orientale itself is on the limb, and with favorable libration and lighting the Rook Mountain inner ring as well as the central mare are observable inside the Cordillera Mountains. Orientale ejecta cover almost everything that is not mare. The Orientale Basin is shown on Chart B8. Grimaldi is also a basin, with the outer ring visible east of Rocca.

Hansteen (F5, 45 km), the **Arrow** (31 km) and **Billy** (G6, 46 km): Similar sized craters with different types of volcanic floors. Billy has a smooth mare-cover, and Hansteen has concentric ridges and cracks, signifying that it is floor-fractured. The Arrow, officially called Mt. Hansteen, is the bright volcanic mass that contains more silica than in mare lavas; it is similar to the Helmet (Chart 23). LO4 image.

Riccioli (A2, 156 km): Large old crater almost completely covered by Orientale ejecta, and later crossed by rilles and small leakage of mare lava on floor. LO4 image.

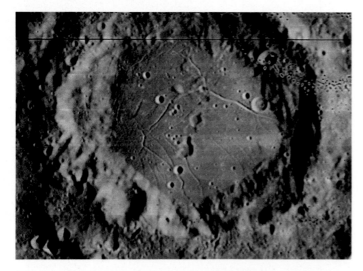

Mersenius (G9, 84 km): Another floor-fractured crater with a domed floor cut by very narrow rilles. The Mersenius Rille to the east, like the Sirsalis Rille, is another rille radial to the putative Procellarum Basin.

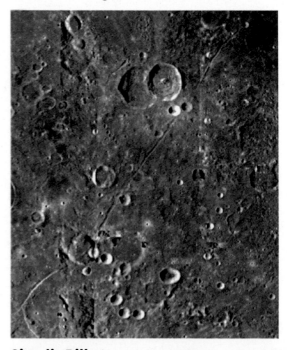

Sirsalis Rille (D5-D8, 405 km): The longest rille (and strongest magnetic anomaly) on the Moon is radial to the proposed Procellarum Basin. Under the rille is the dike that transported lavas to the mare and created the magnetic anomaly.

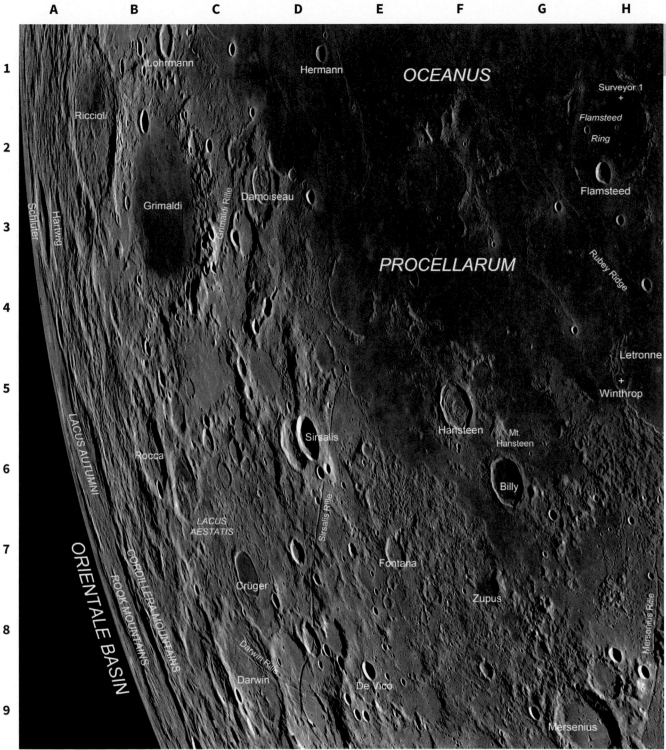

A **B** **C** **D** **E** **F** **G** **H**

1 Lohrmann Hermann *OCEANUS* Surveyor 1 +

Riccioli *Flamsteed*

2 *Ring*

Grimaldi Damoiseau Flamsteed

3 Schlüter Hartwig *PROCELLARUM* Rubey Ridge

4

5 Letronne + Winthrop

Hansteen Mt. Hansteen

6 Rocca Sirsalis Billy

7 *LACUS AESTATIS* Fontana

8 *LACUS AUTUMNI* Crüger Zupus

Darwin Rille Merserius Rille

9 Darwin De Vico Mersenius

CORDILLERA MOUNTAINS *ROOK MOUNTAINS* *ORIENTALE BASIN*

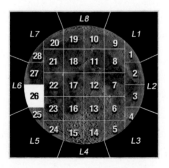

Grimaldi 26

Lunar 100:

L36 - Grimaldi (B3) L68 - Flamsteed Ring (H2)

L44 - Mersenius (G9) L77 - Sirsalis Rille (D5-D8)

L52 - Crüger (C7)

Oceanus Procellarum lavas cover most of this region whose chief areas of interest are the Marius Hills volcanic complex and the Reiner Gamma bright swirl that partly covers the cones. Hedin and Hevelius are veneered by Orientale ejecta – rough and smooth, as are many of the craters near the limb.

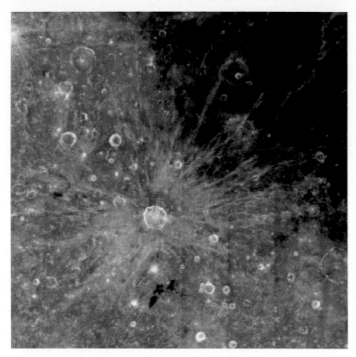

Glushko (A6, 40 km): Very bright oblique impact crater with rays indicating projectile came from the NNW or SSE. Clementine.

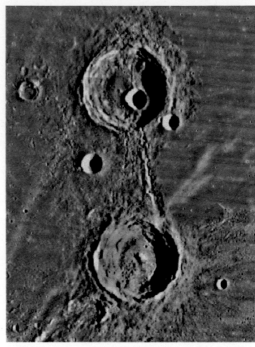

Kraft Crater Chain (B3, 55 km): Probable secondary chain but what is the primary crater? Mare lavas to east cover ejecta from Kraft (top, 51 km); thus are younger. Cardanus (lower 50 km crater) is draped by Glusko ray. LO4 image.

Marius Hills (E4, 235 km): A field of ~300 volcanic cones, cut by small sinuous rilles. Why is this largest concentration of volcanic vents on Moon in central Procellarum?

Reiner Gamma (D6, 460 km): This strangest bright marking on the Moon – a swirl – is visible under high Sun and has no thickness. Entire feature extends from northern tail (about 180 km long), through the central oval (about 40 km) to the twisted splotchy drops of brightness to the south (about 240 km). Associated with a strong magnetic field that perhaps keeps solar wind from darkening soil, but what explains the oval shape and long tails, and why is normally dark mare so bright to begin with? Why does a magnetic field have such a weird shape? Clementine.

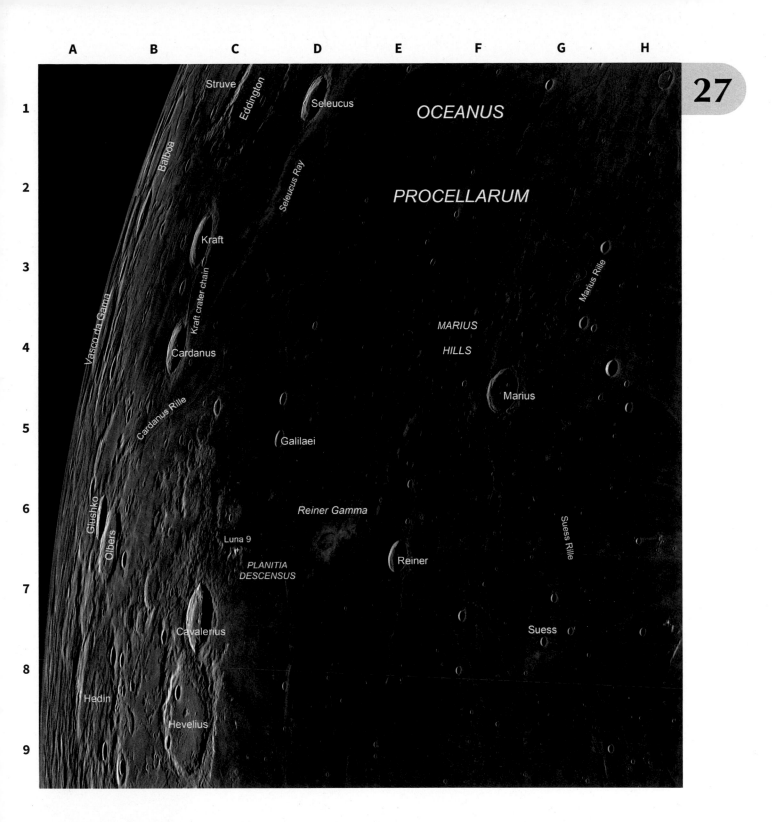

Procellarum 27

Lunar 100:

L42 - Marius Hills (E4)
L57 - Reiner Gamma (D6)

Northern Oceanus Procellarum lavas blandly cover most of this chart with the highlands of the farside just visible along the limb. One of the Moon's most unique landforms, the Aristarchus Plateau, anchors the southeast corner. The sharply bounded plateau appears to have been uplifted and its straight edges suggest fracture boundaries. Schröter's Valley is the largest lunar rille and fed lavas to Procellarum; its Cobra Head vent was probably the source of the very pale mustard-colored pyroclastic deposit that drapes the plateau.

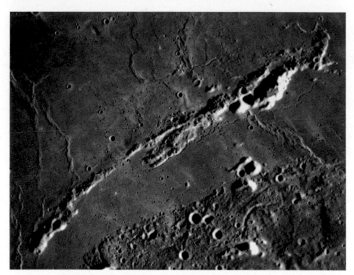

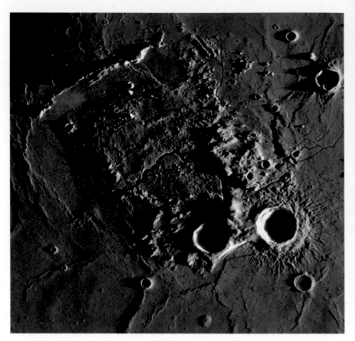

Agricola Mountains (G7, 161 km): The Moon's narrowest mountain range must somehow be associated with the nearby Aristarchus Plateau but no one knows exactly how. It is more mountainous than the Plateau so is not simply a disconnected piece. Apollo 15.

Aristarchus Plateau (H8, 170 x 200 km): The most unusual structural feature on the Moon is this elevated diamond-shaped block dusted by pyroclastics and cut by a huge rille, Schröter's Valley. The Plateau is the source for the Aristarchus Rilles (Chart 21) and younger lavas of northern Oceanus Procellarum.

Aristarchus (H9, 40 km): The brightest crater on the Moon is about ~175 million years old, with well-defined terraces and extensive rays toward Copernicus. Dark and light bands on inner walls previously thought to be changing are not, they are impact melts and avalanche deposits from the rim crest. LO5 image.

Schröter's Valley (H8, 185 km): The Moon's largest sinuous rille, about 6 km wide tapering to 3 km, with a tiny rille on its floor. It was a probable major source of northern Oceanus Procellarum lavas. The Cobra Head, a steep-sided volcanic depression near the summit of a 2 km tall mountain is the source of Schröter's Valley and the location of unlikely red spot "eruptions" reported in 1963.

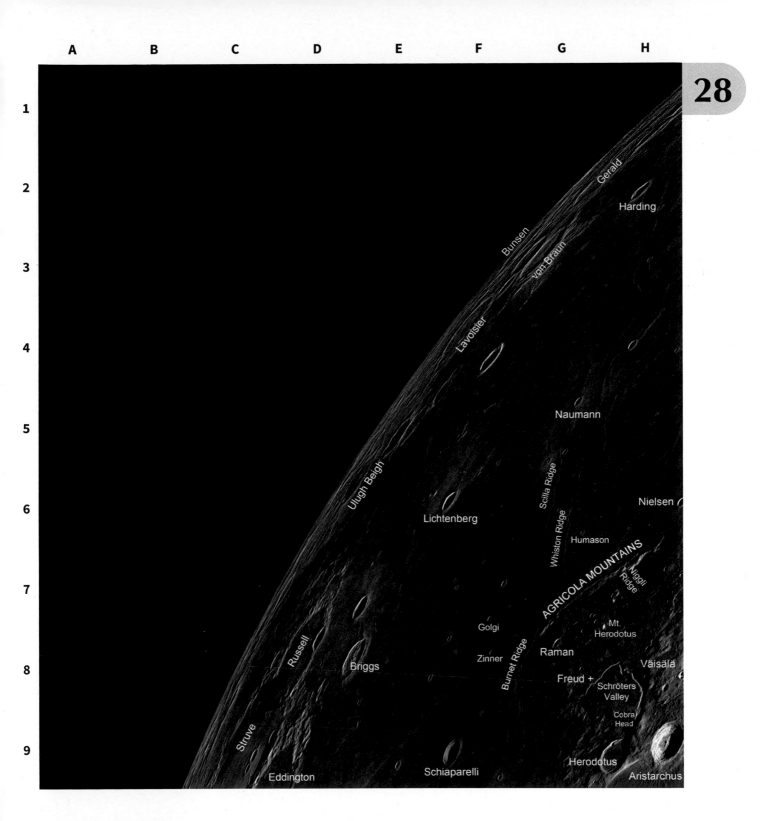

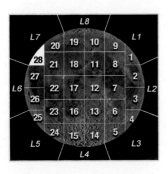

Aristarchus Plateau 28

Lunar 100:

L11 - Aristarchus (H9)
L17 - Schröter's Valley (H8)
L22 - Aristarchus Plateau (H8)

The Limb

Because the Moon is approximately a sphere, circular craters look progressively more oval as its visible edge, or limb, is approached. The main atlas charts display the limb regions as seen from Earth, faithfully reproducing the foreshortening of detail. The following 8 charts are overhead views of the limb that remove the foreshortening, making it easier to understand the geography or selenography of these areas where the nearside and farside meet.

The limb images were made using Jim Mosher's LTVT software to re-project LRO WAC mosaics to provide an overhead view. Nearly all of the named craters are indentified but some smaller ones are omitted in areas of crowding (especially the eastern limb and north pole areas) where craters were named that shouldn't have been. This index map shows the approximate nearside coverage of each sheet which also extends about 20° into the farside. Librations of 8° to 10° bring the edges of the farside into view from Earth, with more distant farside peaks sometimes visible.

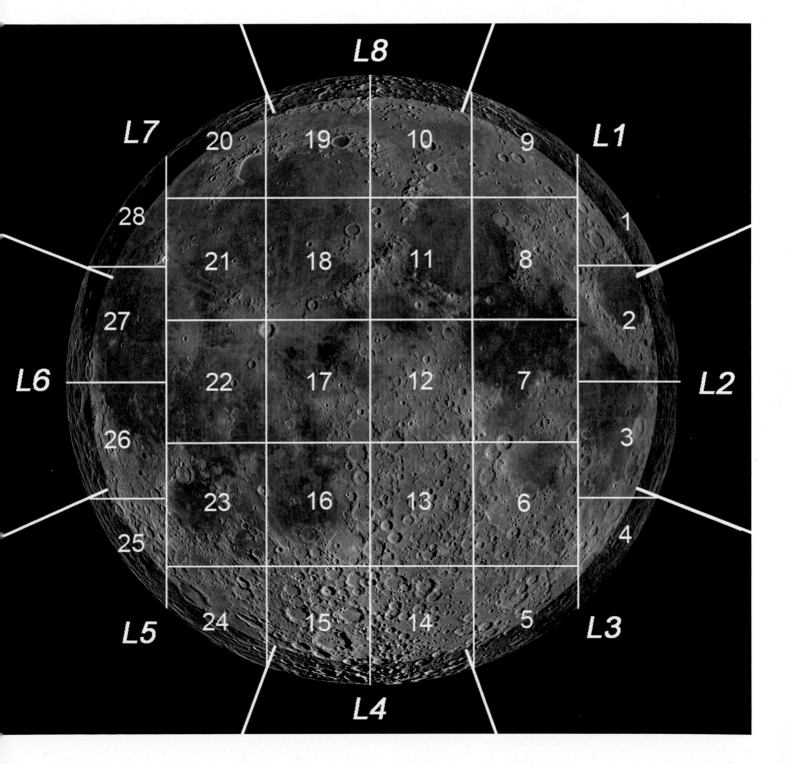

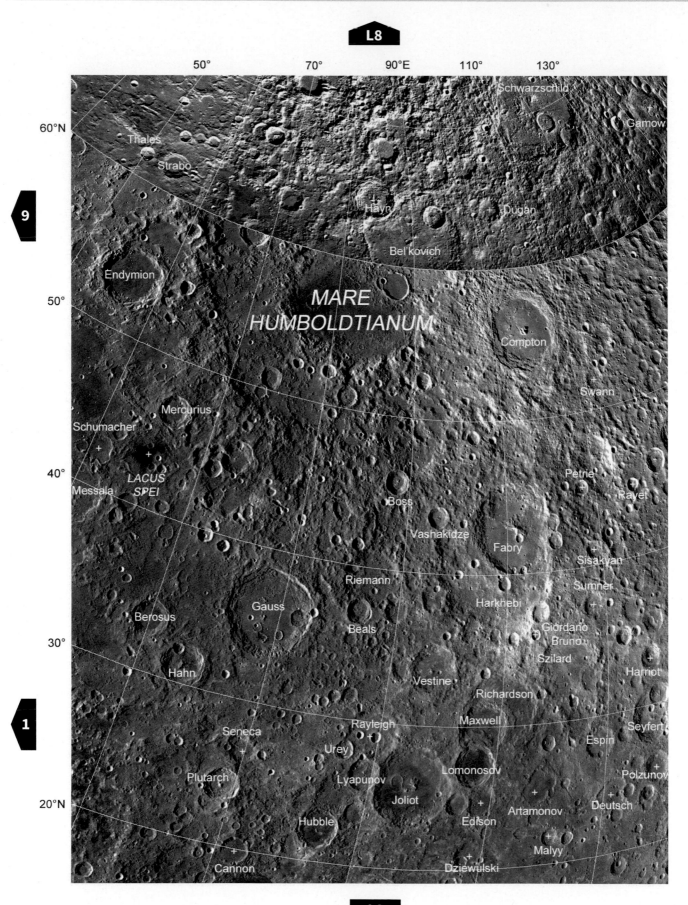

L8

50°　　70°　　90°E　　110°　　130°

60°N

9

Schwarzschild

Gamow

Thales

Strabo

Hayn

Dugan

Bel'kovich

Endymion

50°

MARE
HUMBOLDTIANUM

Compton

Swann

Mercurius

Schumacher

40°

LACUS
SPEI

Petrie

Rayet

Messala

Boss

Vashakidze

Fabry

Sisakyan

Riemann

Sumner

Berosus

Gauss

Beals

Harkhebi

Giordano
Bruno

Szilard

Harriot

Hahn

30°

Vestine

Richardson

Seneca

Rayleigh

Maxwell

Seyfert

Urey

Espin

1

Plutarch

Lyapunov

Lomonosov

Polzunov

Joliot

Artamonov

Deutsch

20°N

Hubble

Edison

Malyy

Cannon

Dziewulski

L2

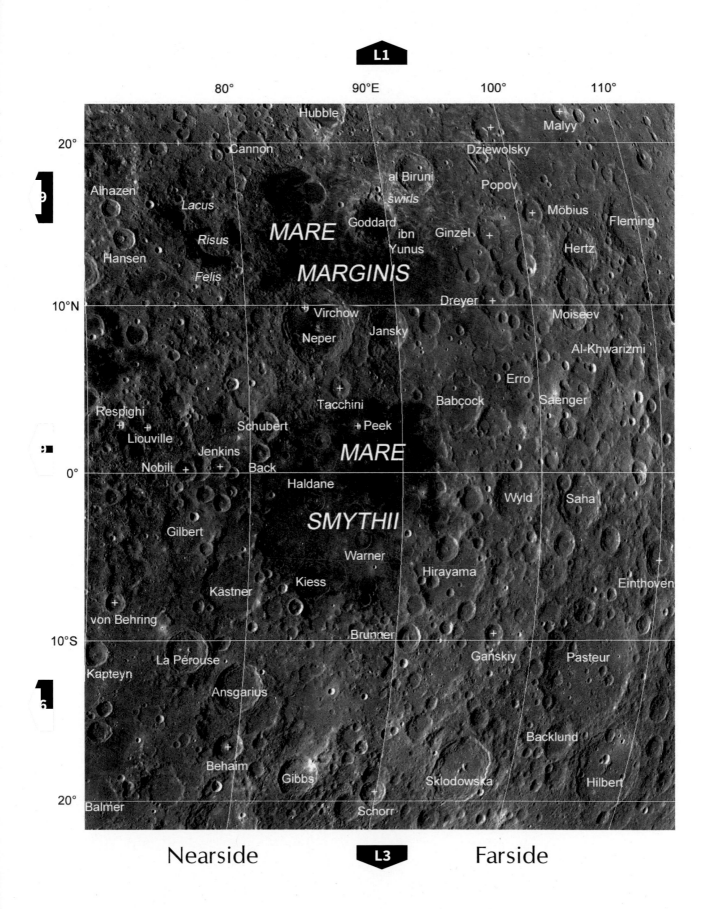

L1

80° 90°E 100° 110°

Hubble

20° Malyy
Cannon Dziewolsky
Alhazen al Biruni
Lacus swirls Popov
Risus Goddard Ginzel Möbius Fleming
ibn
Hansen Yunus Hertz
Felis MARE
MARGINIS
Dreyer
10°N Virchow Moiseev
Neper Jansky
Al-Khwarizmi
Erro
Tacchini Babcock Saenger
Respighi
Liouville Schubert Peek
Jenkins MARE
Nobili Back
0° Haldane
Wyld Saha
Gilbert SMYTHII
Warner
Kiess Hirayama
Einthoven
Kästner
von Behring Brunner
10°S
La Pérouse Ganskiy Pasteur
Kapteyn
Ansgarius
Backlund
Behaim
Gibbs Sklodowska Hilbert
20° Balmer Schorr

Nearside L3 Farside

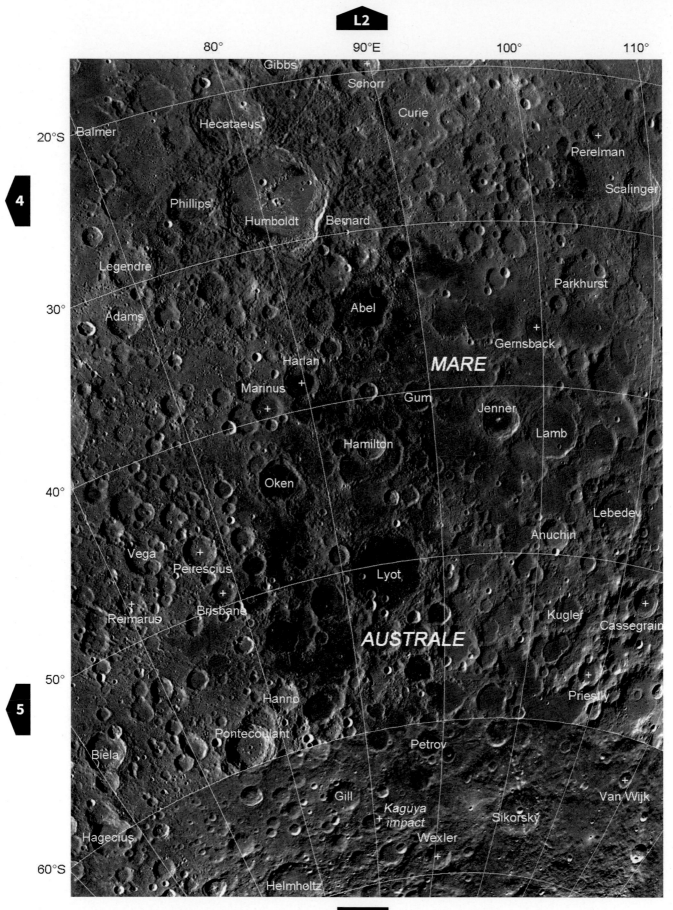

L2

80°   90°E   100°   110°

Gibbs

Schorr

Curie

20°S   Balmer   Hecataeus

Perelman

4

Scalinger

Phillips

Humboldt   Bernard

Legendre

Parkhurst

30°   Adams

Abel

Gernsback

Harlan

MARE

Marinus

Gum   Jenner

Lamb

Hamilton

Oken

40°

Lebedev

Anuchin

Vega

Peirescius   Lyot

Kugler   Cassegrain

Reimarus Brisbane

AUSTRALE

Priestly

50°

5

Hanno

Pontécoulant

Petrov

Biela

Van Wijk

Gill

Kaguya impact   Sikorsky

Hagecius

Wexler

60°S

Helmholtz

L4

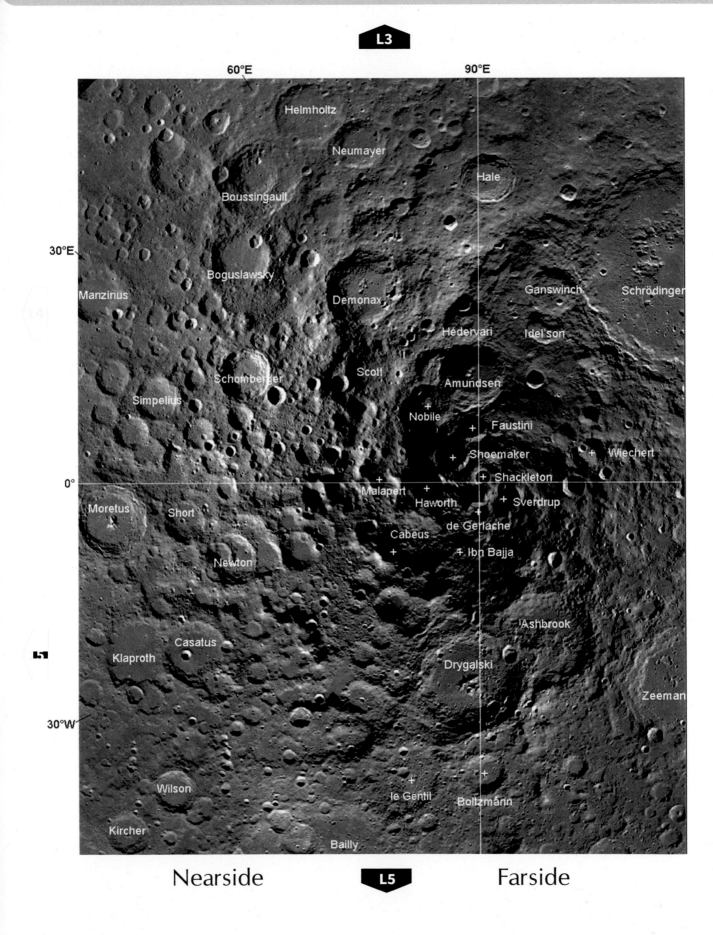

L3

60°E

90°E

Helmholtz

Neumayer

Hale

Boussingault

30°E

Boguslawsky

Ganswinch

Schrödinger

Manzinus

Demonax

Hédervari

Idel'son

Schomberger

Scott

Amundsen

Simpelius

Nobile

Faustini

Shoemaker

Wiechert

Shackleton

0°

Malapert

Sverdrup

Moretus

Short

Haworth

Cabeus

de Gerlache

Newton

Ibn Bajja

Casatus

Ashbrook

Klaproth

Drygalski

Zeeman

30°W

Wilson

le Gentil

Boltzmann

Kircher

Bailly

Nearside   L5   Farside

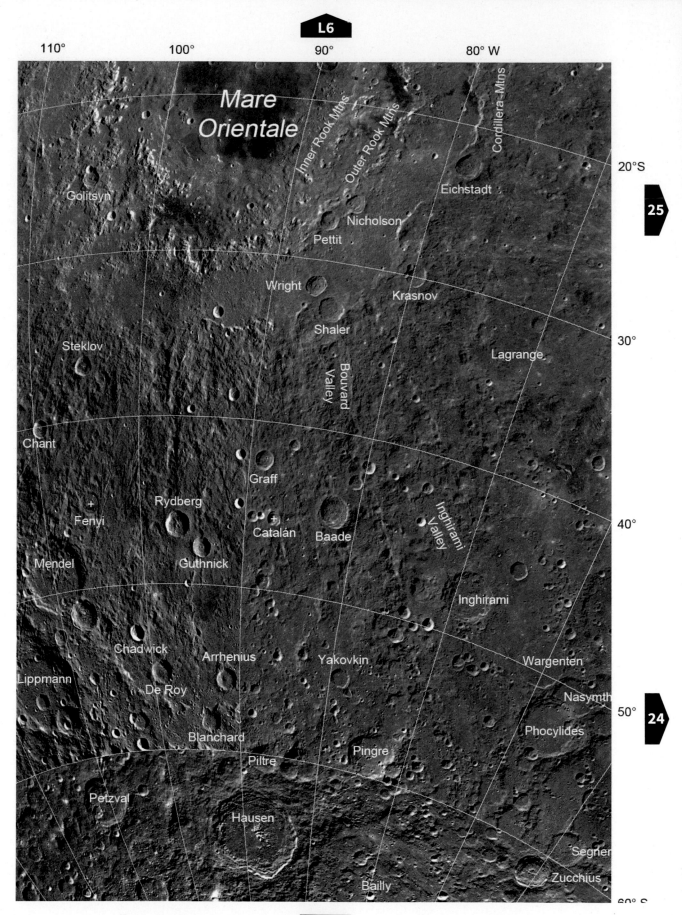

L6

110°   100°   90°   80° W

Mare
Orientale

Inner Rook Mtns
Outer Rook Mtns
Cordillera Mtns

Golitsyn

20°S

25

Eichstadt

Nicholson
Pettit

Wright

Krasnov

30°

Shaler

Lagrange

Steklov

Bouvard
Valley

Chant

Graff

Rydberg

Inghirami
Valley

40°

Fenyi
Catalán   Baade

Mendel   Guthnick

Inghirami

Chadwick   Arrhenius   Yakovkin

Wargenten

Lippmann   De Roy

Nasymth

50°   24

Blanchard   Phocylides

Piltre   Pingre

Petzval

Hausen   Segner

Bailly   Zucchius

60° S

L4

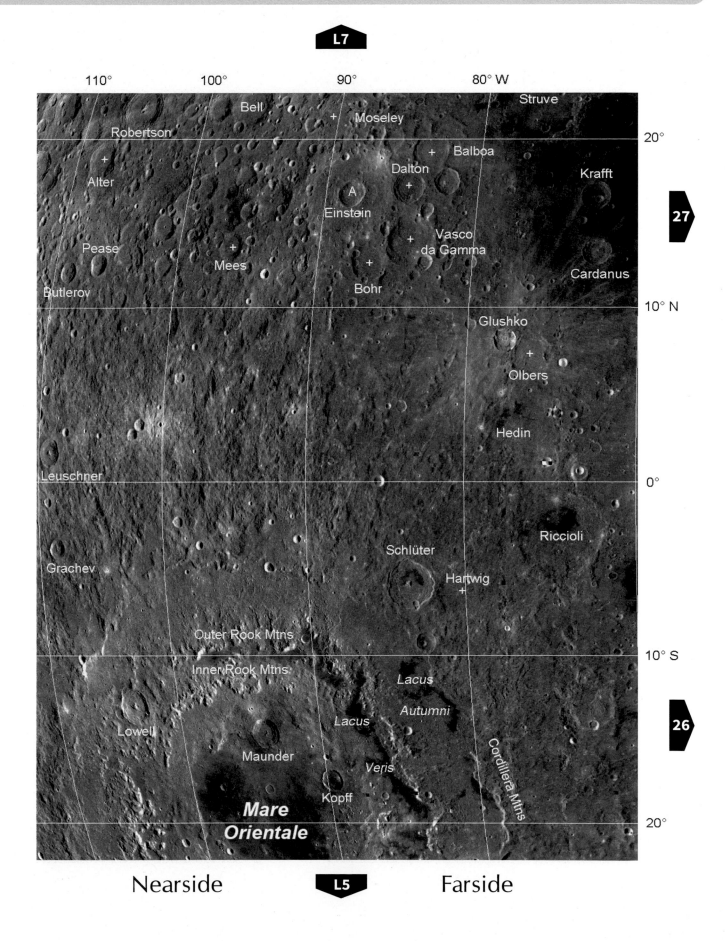

L7

| 110° | 100° | 90° | 80° W |

Struve

Bell

Moseley

+

Robertson

20°

+ Balboa

Alter

+

Dalton

Krafft

A

+

Einstein

Vasco
da Gamma

+

Pease

+

Mees

Bohr

+

Cardanus

Butlerov

10° N

Glushko

+

Olbers

Hedin

Leuschner

0°

Riccioli

Schlüter

Grachev

Hartwig

+

Outer Rook Mtns

10° S

Inner Rook Mtns

Lacus

Autumni

Lacus

Lowell

26

Maunder

Veris

Cordillera Mtns

Kopff

20°

**Mare
Orientale**

Nearside   L5   Farside

27

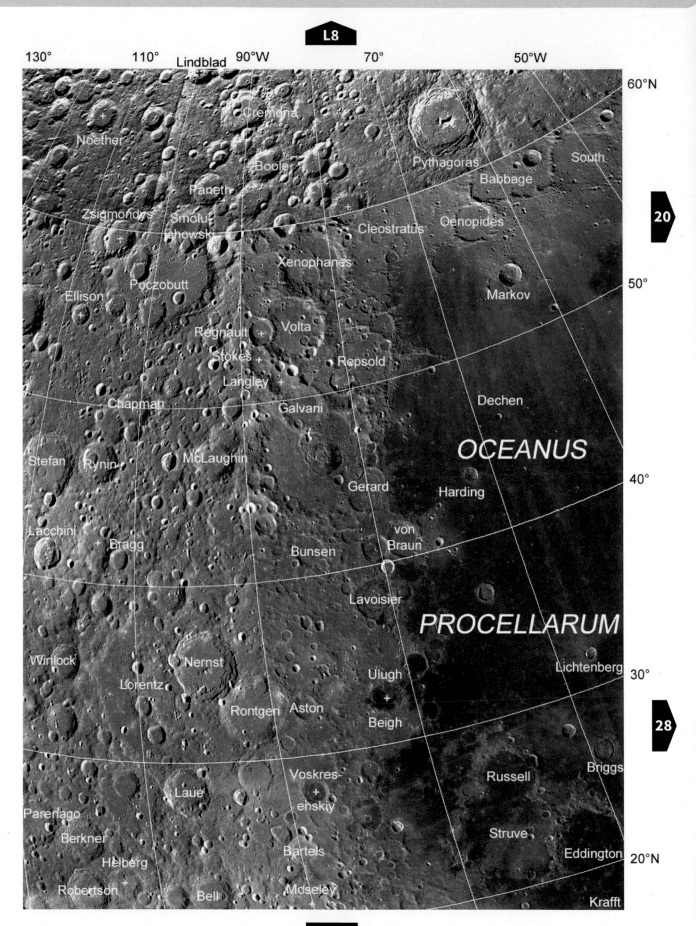

L8

130°   110°   Lindblad   90°W   70°   50°W   60°N

Noether   Cremona

Boole   Pythagoras   South

Paneth   Babbage

Zsigmondys   Smolu-   Cleostratus   Oenopides   20

Smoluchowski

Xenophanes   50°

Ellison   Poczobutt   Markov

Regnault   Volta

Stokes   Repsold

Langley   Dechen

Chapman   Galvani   OCEANUS   40°

Stefan   Rynin   McLaughin

Gerard   Harding

Lacchini   von   Braun

Bragg   Bunsen   PROCELLARUM

Lavoisier

Winlock   Nernst   Lichtenberg   30°

Lorentz   Ulugh

Rontgen   Aston   Beigh   28

Voskres-   Briggs

Laue   enskiy   Russell

Parenago

Berkner   Struve

Helberg   Eddington   20°N

Robertson   Bell   Moseley   Krafft

L6

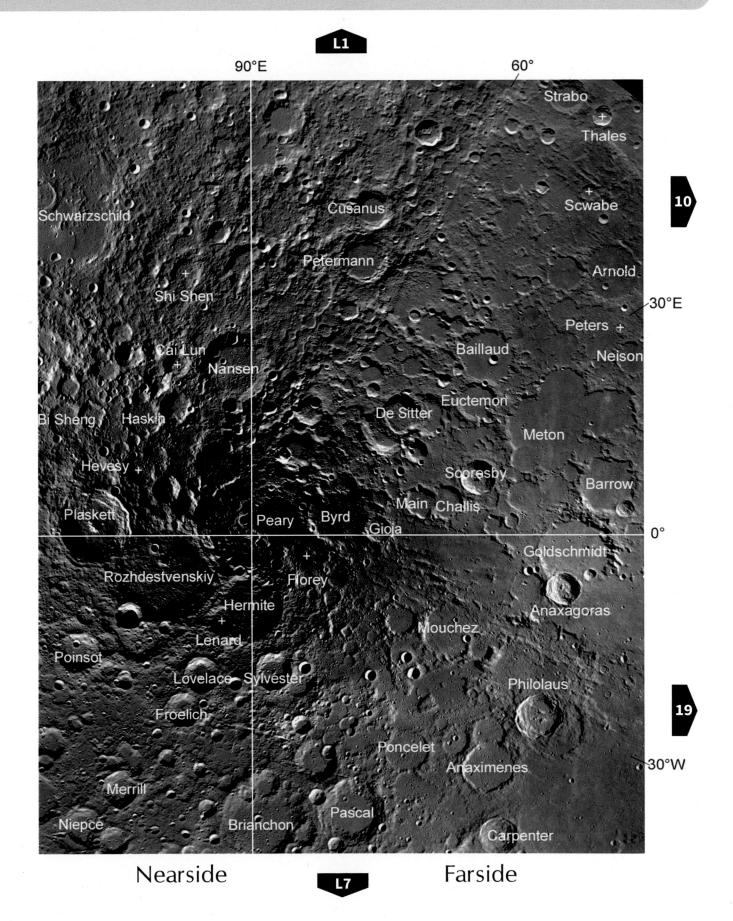

L1

90°E   60°

Strabo
Thales +

10

Schwarzschild

Cusanus   Scwabe +

Petermann   Arnold

Shi Shen +   30°E

                                              Peters +
Cai Lun +   Baillaud   Neison
        Nansen

Bi Sheng   Haskin   De Sitter   Euctemon

                                              Meton
Hevesy +

                              Scoresby   Barrow
Plaskett
        Peary   Byrd   Main Challis
                    Gioja   0°

                                        Goldschmidt
Rozhdestvenskiy +
              Florey   Anaxagoras
        Hermite +
        Lenard   Mouchez
Poinsot
        Lovelace Sylvester   Philolaus   19

Froelich
                            Poncelet   30°W
Merrill   Anaximenes
Niepce   Brianchon   Pascal
                                      Carpenter

Nearside   L7   Farside

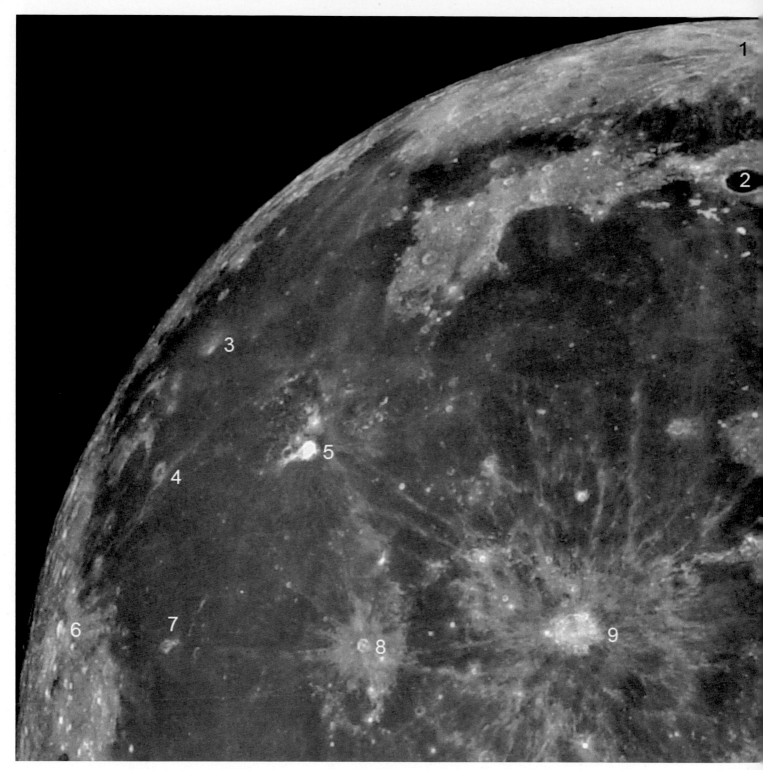

Consolidated Lunar Atlas Yerkes Y842

1: Anaxagoras (Chart 19)	4: Seleucus (28)	7: Reiner Gamma (27)
2: Plato (9)	5: Aristarchus (28)	8: Kepler (22)
3: Lichtenberg (28)	6: Glushko (27)	9: Copernicus (22)

Consolidated Lunar Atlas Yerkes Y842

1: Anaxagoras (19)	4: Atlas (9)	7: Manilius (11)
2: Thales (9)	5: Geminus (1)	8. Proclus (2)
3: Eudoxus (10)	6: Menelaus (11)	9: Dionysius (12)

Consolidated Lunar Atlas Yerkes Y842

1: Grimaldi (26)	4: Byrgius A (25)	7: Schickard (24)
2: Lalande (17)	5: Bullialdus (23)	8: Tycho (15)
3: Gassendi (23)	6: Doppelmayer pyroclastics (23)	9: Zucchius (24)

Consolidated Lunar Atlas Yerkes Y842

1: Alphonsus (16)	4: Mädler (7)	7: Cassini's Bright Spot
2: Hipparchus C (12)	5: Langrenus (3)	8: Werner (13)
3: Cyrillus A (6)	6: Petavius B (3)	9: The Headlights (4)

Mare Humboldtianur

B = Bel'kovich
C = Compton
E = Endymion

Mare Ridges
None named

A two-ringed impact basin with a hint of a concentric mare ridge on the west and southwest. The prominent outer ring informally called the Andes Mountains is 650 km in diameter, and the inner ring informally named the Bishop Mountains, make a circle 340 km in diameter.

Mare Marginis

G = Goddard
H = Hubble
N = Neper

Mare Ridges
None named

A mare with no obvious underlying basin. Only one arc of a mare ridge curves from south of Goddard toward Hubble.

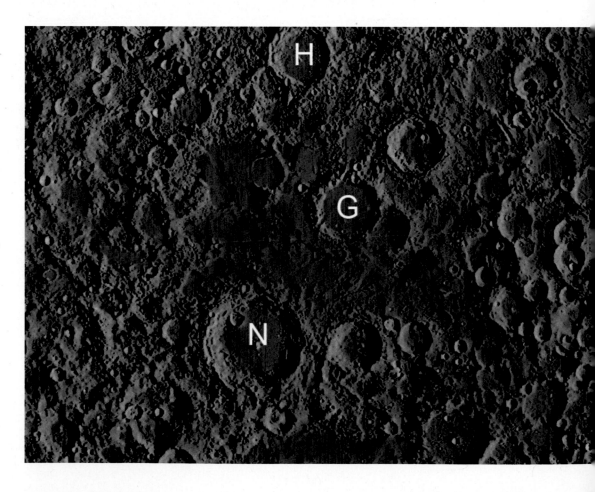

Mare Crisium

C = Cleomedes
F = Firmicus
M = Macrobius

Mare Ridges

1 = Oppel
2 = Tetyaev
3 = Termier
4 = Harker

W-E elongated basin (due to an oblique impact) with massive 740 km diameter main rim and mare ridges inner ring. Mare level steps down ~500 m inside mare ridge ring. An outer 1080 km wide ring curves between Hahn and Geminus (Chart 1).

Mare Smythii

H = Hirayama
K = Kästner
N = Neper

Mare Ridges

1 = Dana
2 = Cloos

Strong basin rim on west with inner basin ring hinted at by curved mare ridge ring on west of floor. Mare bounding ring is 540 km in diameter.

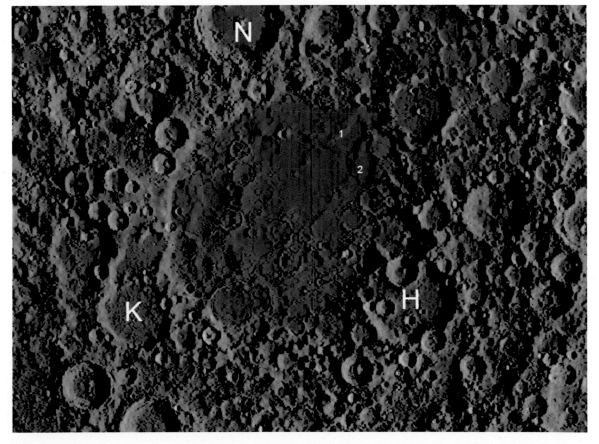

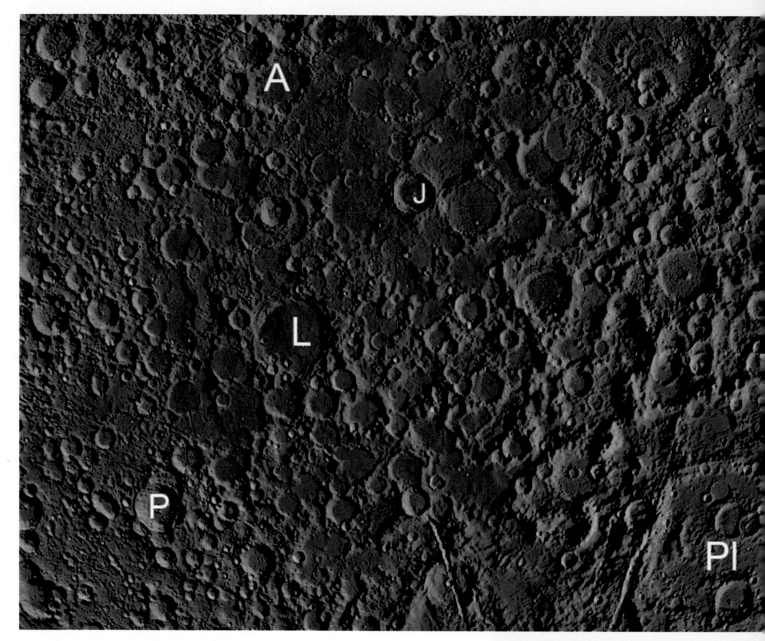

Mare Australe

Australe is an ancient impact basin, detectable now as a circular area where craters and spaces between them are filled with mare lava. Two uncertain rings suggest an initial basin 880 km in diameter, and altimetry shows that the center is about 2 km lower than surrounding terrain. An unnamed mare ridge crosses Lyot but a larger and more unusual one cuts Oken (Chart L3) and mare outside the crater to the northwest of Lyot. Straight chains/gouges near Planck radiate from Schrödinger (Chart L4).

A = Abel
J = Jenner
L = Lyot
P = Pontecoulant
Pl = Planck Basin

Mare Ridges
None named

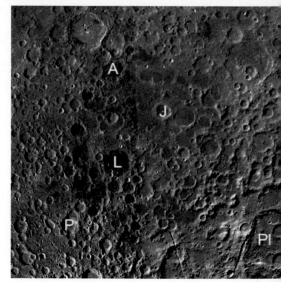

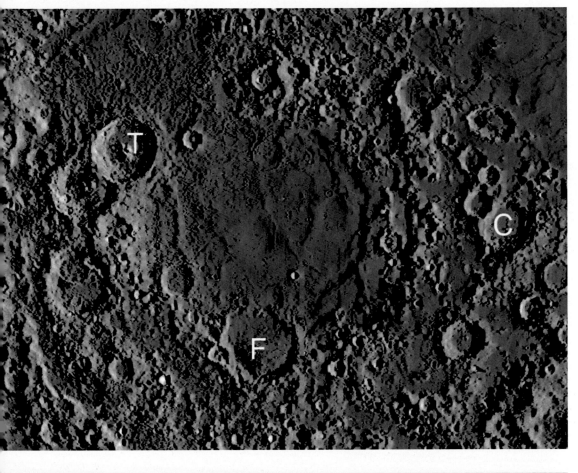

Mare Nectaris

C = Columbo
F = Fracastorius
T = Theophilus

Mare Ridges
None named

Classic basin with Altai Scarp 860 km wide main rim and second and third rings passing through Catharina and Fracastorius. Strong mare ridge ring on east defines a smaller inner ring.

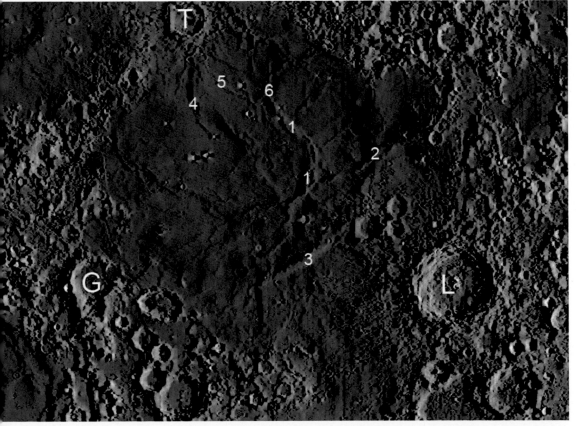

Mare Fecunditatis

G = Gutenberg
L = Langrenus
T = Taruntius

Mare Ridges
1 = Geikie
2 = Andrusov
3 = Mawson
4 = Cato
5 = Cushman
6 = Cayeux

Western half of mare is topographically low and perhaps a basin. East half high and ridgy; it is uncertain if a basin lurks beneath the lava.

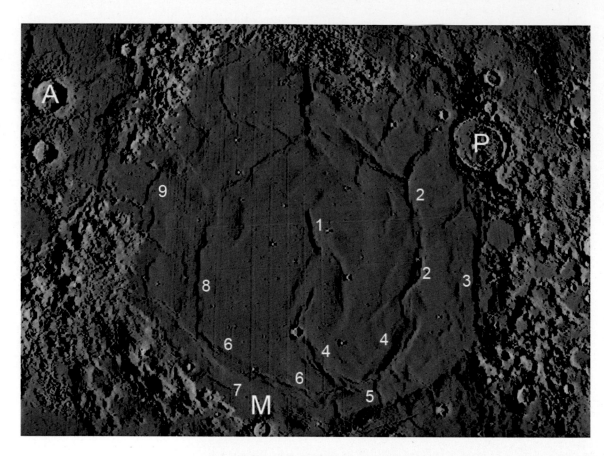

Mare Serenitati.

A = Aristillus
M = Menelaus
P = Posidonius

Mare Ridges
1 = Azara
2 = Smirnov
3 = Aldrovandi
4 = Lister
5 = Nicol
6 = Buckland
7 = Sorby
8 = Von Cotta
9 = Gast

Ridges 2 and east part of 4 traditionally called the Serpentine Ridge. Haemus Mountain basin ring is about 920 km in diameter. Smaller basin at upper left.

Mare Tranquillitatis

A = Arago
L = Lamont
M = Maskelyne
P = Plinius

Mare Ridges
1 = Barlow

Western half is ringed with concentric rilles and includes massive curved mare ridge cut by Lamont. Eastern half of mare is high and almost ridgeless, with surface fractured by Imbrium-radial Cauchy Rille and Fault.

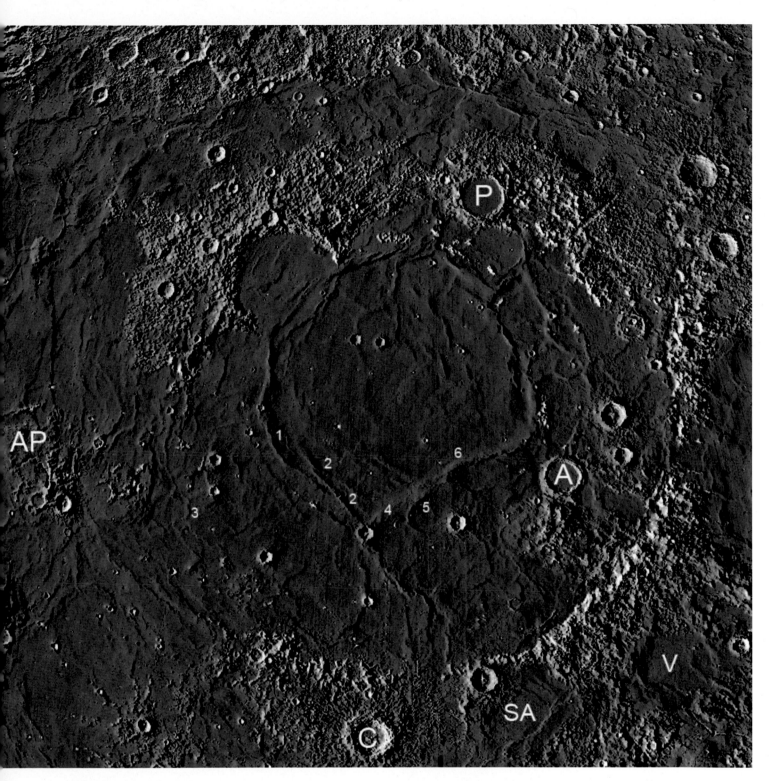

Mare Imbrium

Magnificent multi-ring basin with 1160 km diameter Apennine main rim, a ring passing though Archimedes, and an inner one defined by a circular mare ridge system with names only for southern half. Where is the northern side of the Apennine rim? Is it the north shore of Frigoris? See back cover for an overhead view.

A = Archimedes
AP = Aristarchus Plateau
C = Copernicus
P= Plato
SA = Sinus Aestuum
V = Mare Vaporum

Mare Ridges
Mare Ridges
1 = Heim
2 = Zirkel
3 = Arduino
4 = Stille
5 = Higazy
6 = Grabau

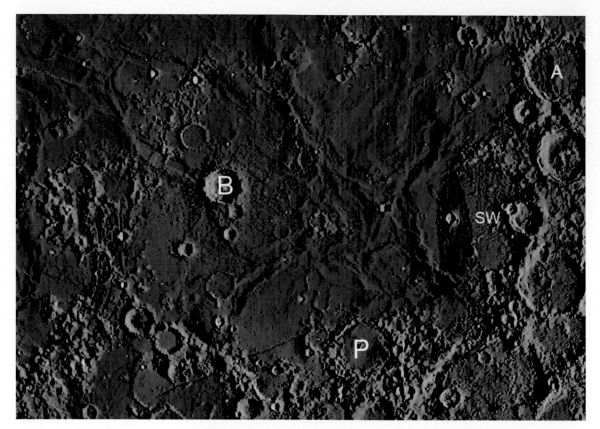

Mare Nubium

A = Alphonsus
B = Bullialdus
P = Pitatus
SW = Straight Wall

Mare Ridges
None named

A 690 km wide basin is indicated by partial mountain rim to southwest and a topographic low. Is the curved ridge under Bullialdus an inner basin ring? Notice the downward slope from the Straight Wall towards the west.

Mare Humorum

G = Gassendi
M = Mersenius
V = Vitello

Mare ridges
None named

An old basin with pieces of its main, 425 km diameter rim visible between Gassendi and Mersenius and Vitello. Mare ridges define an inner basin ring, and an outer one is just beyond Mersenius.

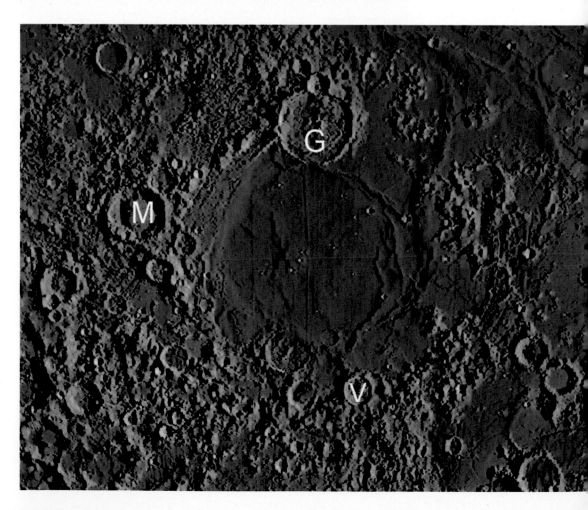

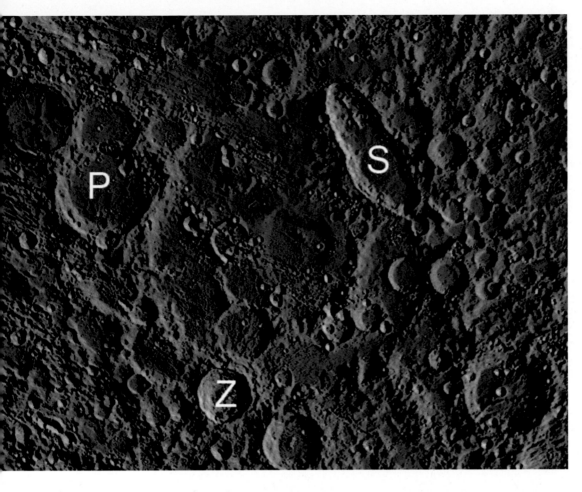

Schiller-Zucchius Basin

P = Phocylides
S = Schiller
Z = Zucchius

Mare Ridges
None named

Not discovered until the 1960s the SZB has outer (335 km) and inner (175 km) rims visible to the south, and missing to the north. This topo image confirms a third inner ring or depression.

Mare Orientale

L = Lowell
M = Maunder
S = Schlüter
W = Wright

Mare ridges
None named

The Moon's youngest large basin has the magnificent Cordillera (930 km) and Outer Rook (620 km) rings but uncertain inner rings hinted at by ridges. The lack of deep mare fill suggests what the Imbrium Basin looked like before lavas flooded its floor.

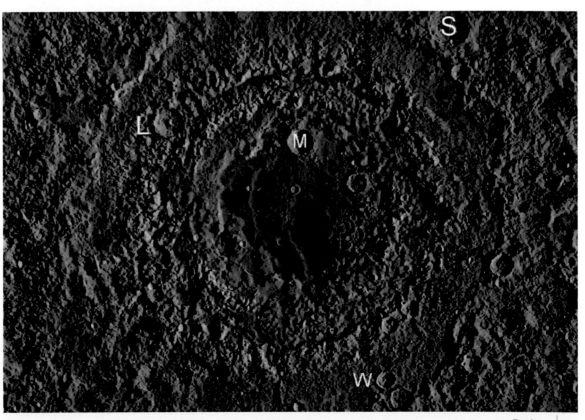

Oceanus Procellarum

AP = Aristarchus Plateau
MH = Marius Hills
R = Rümker

Mare Ridges
1 = Scilla
2 = Whiston
3 = Burnet
4 = Bucher
5 = Argand
6 = Arduino
7 = Rubey
8 = Ewing

Procellarum is the largest mare on the Moon and a continuing enigma - does it fill an impact basin? Two possible basins have been proposed, the best defined simply being called the Procellarum Basin. The alternative hypothesis is that Procellarum is a low area surrounding the Imbrium Basin, and that Imbrium may have formed by oblique impact. In any case the mare ridges tend to follow the curved shoreline, implying that they have a common structural reason for existing.

The area of the propsed Procellarum Basin contains nearly all of the Moon's radioactive terrain, which geochemists call the Procellarum KREEP Terrain or PKT. KREEP elements are potassium (K), rare earth elements (REE) and phosphorus (P). KREEP formed early in lunar history and exists only in the Procellarum Basin area.

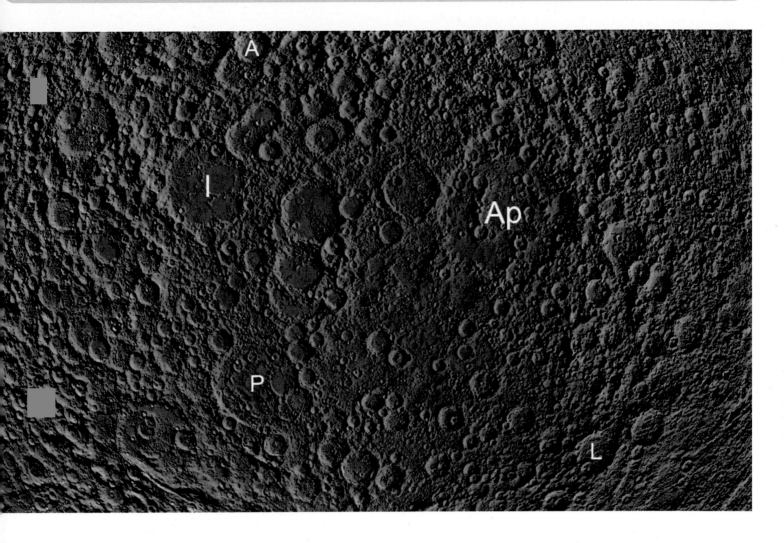

South Pole-Aitken Basin (SPA)

The largest and oldest lunar basin is mostly defined by its great depth (~13 km) and occasional mountain rim segments, especially to the north-northeast and near the South Pole where the once-named Leibnitz Mountains loom for terrestrial observers. In high Sun views (Chart F4) the interior of the basin is dark because of exposed iron-rich rocks, including some mare lavas. Occasional short mare ridges occur on floors of smaller contained basins, but there is little evidence for basin interior rings for SPA. They were undoubtedly formed, but as at the Australe Basin are no longer visible.

A = Aitken
Ap = Apollo Basin
I = Mare Ingenii
L = Lippmann
P = Poincare Basin

Mare Ridges
None named

Landing Sites: Apollo

The most famous lunar landing was Apollo 11, where humans first touched down on the lunar surface. Since Luna 2 hit the Moon in 1959 there have been 18 controlled landings on the surface (Surveyors, Apollos, and Lunas and their Lunakhod cargos) 13 planned crashes (Rangers, early Lunas, Lunar Prospector, LCROSS, Kaguya, Chang'e 1, Chandrayaan 1), and 5 unplanned crashes (Lunas and Surveyors). Many of these landing/crash sites and pieces of spacecraft have been imaged by the Lunar Reconnaissance Orbiter high-resolution camera. No evidence of lunar landings is visible from Earth, but it is fascinating to telescopically identify the locations of the classic landings.

Apollo 11 (Chart 7; S5 = Surveyor 5; R8 = Ranger 8)

Apollo 12 (22; B = Fra Mauro B; K=Montes Riphaeus K)

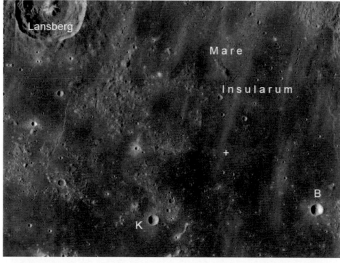

Apollo 14 (17; H & HA = Fra Mauro H & Fra Mauro HA)

Apollo 15 (11; H = Hadley Mtn; HD = Hadley Delta Mtn)

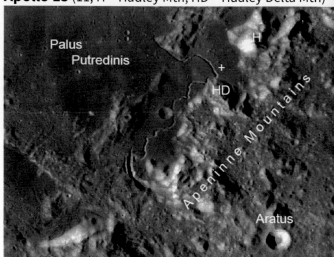

Apollo 16 (Chart 12; NR & SR = North & South Ray crater)

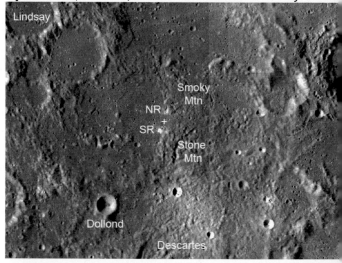

Apollo 17 (Chart 8; N, S, E = North, South, East Massif)

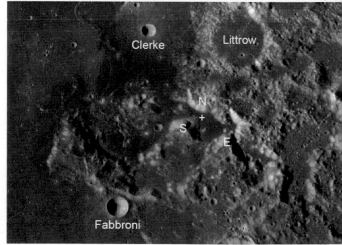

Luna 9 (Chart 26)

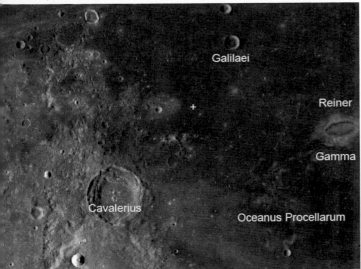

Galilaei

Reiner

Gamma

Cavalerius

Oceanus Procellarum

Surveyor 7 (Chart 15)

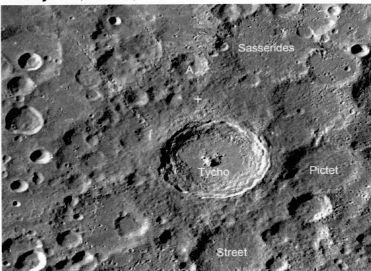

Sasserides

A

Tycho

Pictet

Street

Luna 21 & Lunakhod 2 (8; +=landing site; F=final stop)

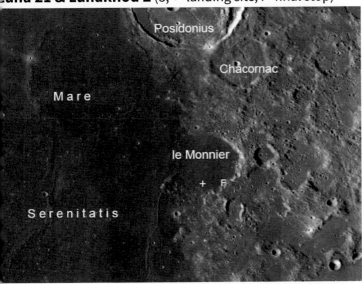

Posidonius

Chacornac

M a r e

le Monnier

+ F

S e r e n i t a t i s

Ranger 9 (Chart 16)

Ptolemaeus

+

Alphonsus

Arzachel

Luna 24 (Chart 2)

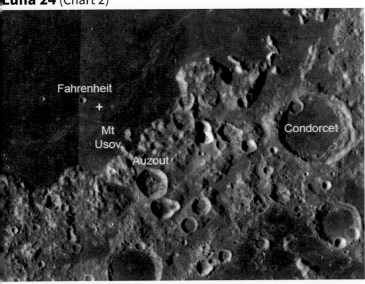

Fahrenheit

+

Mt
Usov

Auzout

Condorcet

LCROSS in Cabeus & Lunar Prospector in Shoemaker- 15

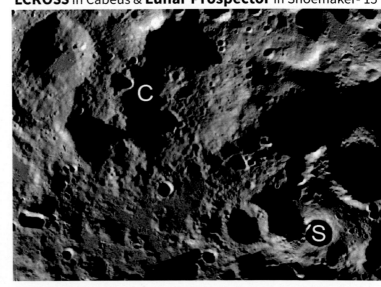

C

S

The Farside

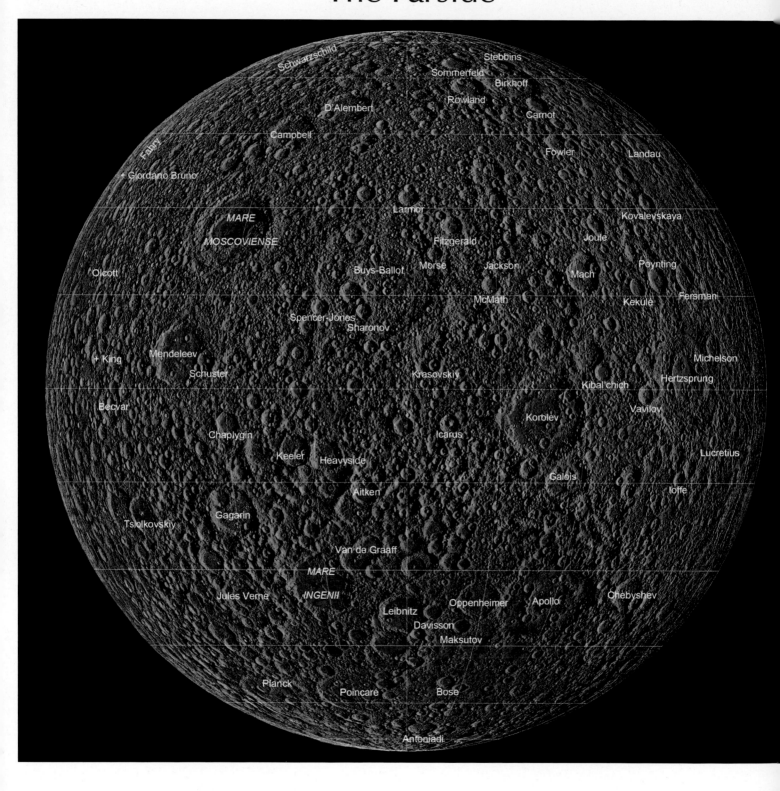

The Farside

If the Moon were not tidally locked – as is every other moon in the solar system – we would have two wildly different views of our satellite. Locked rotation means that the gravitational attraction of the planet gradually balances with the satellite's distance and rotation period to make one side of the satellite always face the planet.

The familiar lunar nearside is heavily cratered in the deep southern hemisphere but about 35% of the near-side is covered with dark mare lavas that fill large lunar basins. The farside has large basins too – including the largest in the solar system – but very little of the mare lavas erupted on to the farside, which hence lacks the albedo differences that divide the nearside into distinct regions. The lack of significant maria also means that fascinating volcanic landforms such as sinuous rilles, domes and mare ridges are uncommon. If synchronous rotation requires that we only get to see one side, we got the best view.

The Big Backside Basin

The farside is a hemisphere of impact craters of all sizes, piled on top of each other. The nearside basins and their mare filling wiped out craters from the early period of intense bombardment but that didn't happen as much on the farside so craters are everywhere. Or almost everywhere, for the southern portion of the far-side is dominated by the South Pole-Aitken Basin (SPA). Its name comes from bounding features - the crater Aitken at 16.8°S and the South Pole. From the nearside we see in profile part of the high rim of SPA as the Leibnitz Mountains (Chart #15).

Once called the Big Backside Basin, this 2500 km wide giant formed sometime in the first 600 million years of lunar history - it hasn't been dated yet. It excavated the deepest hole in the Moon and even now there is 13 km of elevation difference

between its remnant rim and center. Because it cuts down so deeply into the lunar crust, magma apparently was more able to reach the surface so that many of the farside's maria occur in smaller basins located within the SPA.

The structure of the SPA is difficult to see in images such as this Clementine mosaic, but its darker hue is obvious. On a topographic map the basin stands out as the largest feature of the Moon. The most visible part of its rim on the farside is a mountainous 90° arc between the Galois and Chebyshev craters (Chart F4). An inner ring is visible southwest of Galois.

Other Basins

The second largest farside basin is Orientale, which is centered at 92.8°W. The Cordillera and Rook Mountains have the most continuous rings of any large basin, and the farside ejecta of Orientale contains the most spectacular basin secondary crater chains on the Moon (see image p 4 of introduction).

The Apollo Basin is within SPA and contains patches of maria on its floor and in a moat between rings. Small craters within Apollo are named for American astronauts who have died. The nearby Korolov Basin is similarly a nomenclatural cemetery for Soviet and Russian cosmonauts.

Moscovience is a multi-ring basin with a large enough pond of dark lava on its floor to be officially named a maria. A second large exposure of mare outside SPA covers the floor of the 184 km diameter young crater Tsiolkovskiy.

Ray Craters

The farside has many bright ray craters with the most spectacular being Jackson. Like Tycho it has a dark collar of impact melt and was formed by an oblique impact. Jackson's zone of avoidance is very sharply defined by its asymmetric rays, indicating that the projectile came from the northwest at a lower angle than Tycho's projectile. Ohm is another fresh ray crater of 62 km diameter surrounded by a dark collar. Its zone of avoidance implies an oblique impact with a projectile coming from the north-northeast. Ohm's rays extend onto the nearside, reaching north of the Aristarchus Plateau. A third ray feature is really a peculiar complex extending from Necho past King northward to Giordano Bruno. One of Bruno's rays wraps around to the nearside Alhazen crater near Crisium.

Other Remarkable Features

King crater is famous for its claw-like central peak and immense impact melt deposit most visibly ponded in the depression to the north.
Buys-Ballot is a Schiller-like oblique impact crater 90 km long and about 60 km wide. Like Schiller it has a central ridge rather than a central peak.
One of most spectacular lunar swirl fields occurs on the dark floor of Mare Ingenii, which being antipodal to the Imbrium Basin led to suggestions that swirls are related to basin impacts. A new theory is that all swirls are due to meteorite-rich ejecta from SPA.

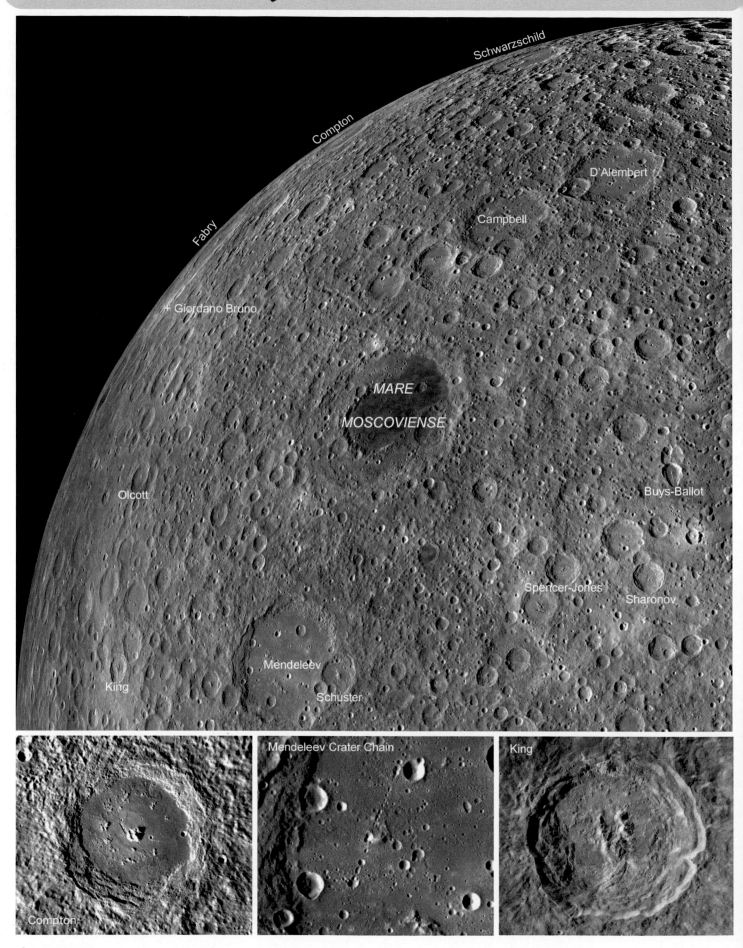

Schwarzschild

Compton

D'Alembert

Campbell

Fabry

+ Giordano Bruno

MARE

MOSCOVIENSE

Buys-Ballot

Olcott

Spencer-Jones

Sharonov

Mendeleev

King

Schuster

Compton

Mendeleev Crater Chain

King

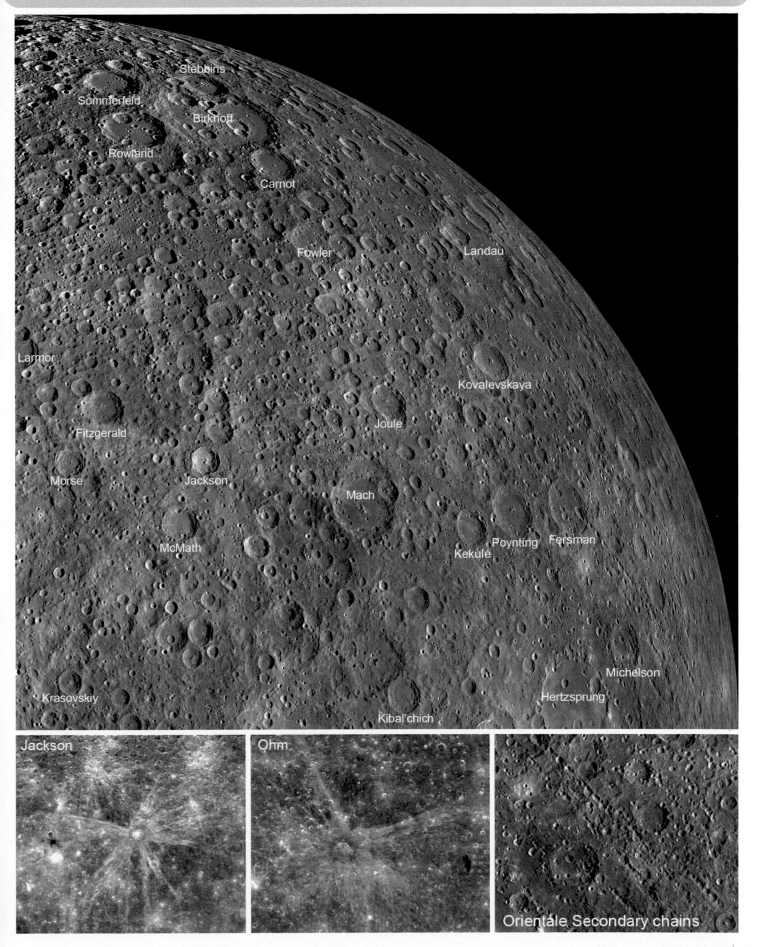

Stebbins

Sommerfeld

Birkhoff

Rowland

Carnot

Fowler

Landau

Larmor

Kovalevskaya

Joule

Fitzgerald

Morse

Jackson

Mach

McMath

Poynting   Fersman

Kekulé

Kibal'chich

Michelson

Krasovskiy

Hertzsprung

Jackson

Ohm

Orientale Secondary chains

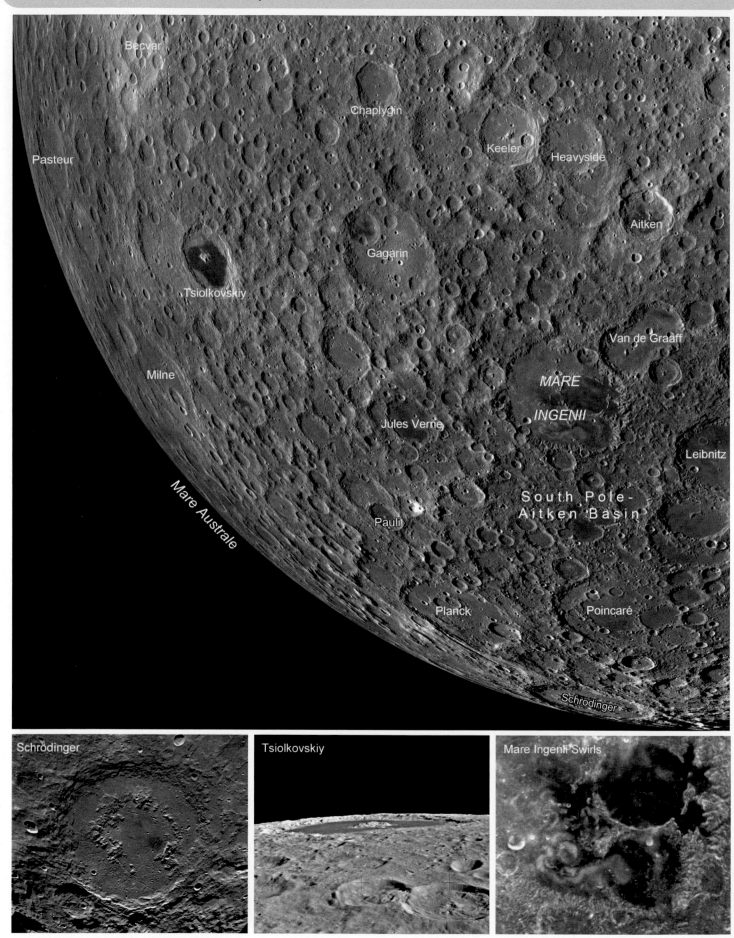

Becvar

Chaplygin

Pasteur

Keeler

Heavyside

Aitken

Gagarin

Tsiolkovskiy

Van de Graaff

Milne

MARE

INGENII

Jules Verne

Leibnitz

Mare Australe

South Pole-
Aitken Basin

Pauli

Planck

Poincaré

Schrödinger

Schrödinger

Tsiolkovskiy

Mare Ingenii Swirls

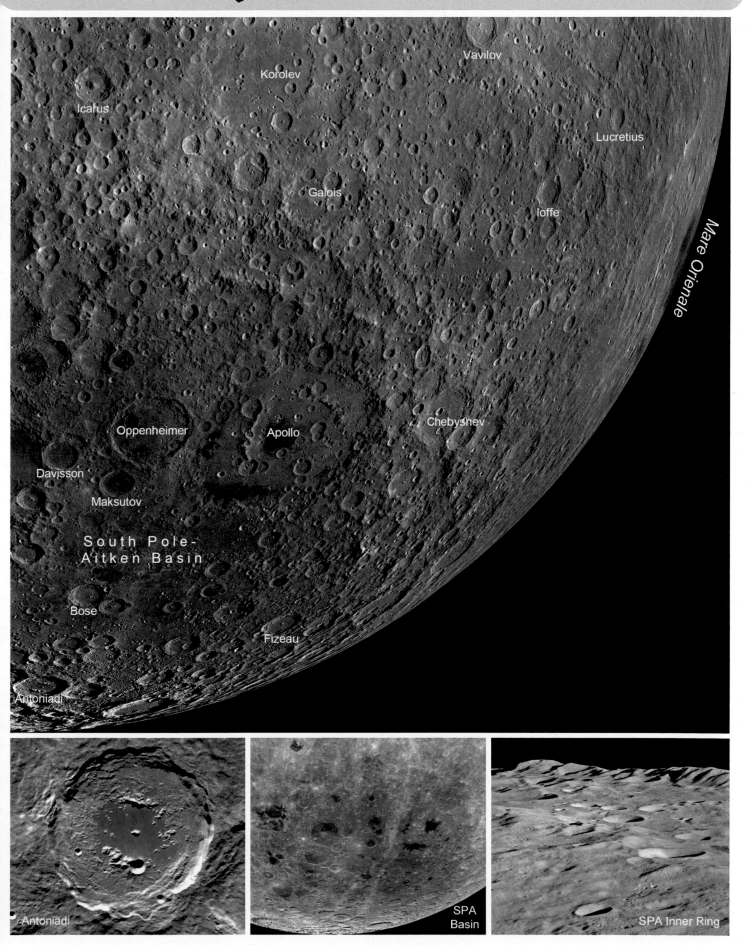

Vavilov

Korolev

Icarus

Lucretius

Galois

Ioffe

Mare Orientale

Chebyshev

Oppenheimer

Apollo

Davisson

Maksutov

South Pole-
Aitken Basin

Bose

Fizeau

Antoniadi

Antoniadi

SPA
Basin

SPA Inner Ring

Errata

P. 28: Image is of Eratosthenes, not Plinius.

P. 32: Cassini image should be rotated 180º.

P. 32: Caucasus Mountains. Replace description with: Caucasus Mountains (C9, 444 km): Part of the Imbrium Basin rim that strangely does not curve around the mare but is straight. Does this imply that the basin rim does not pass near Plato, but perhaps along the north shore of Mare Frigoris? See back cover image.

P. 32, P. 33 and P. 102: "Alpes" should be "Alps" in keeping with English (not Latin) names for mountain ranges.

P. 64: Arrowhead mistakenly labeled Arrow

Typographical errors of lunar nomenclature (correct spelling given)

P. 4: Altai Scarp

P. 5: Mairan T

P. 13: Deslandres

P. 13: Mare Tranquillitatis

P. 15 and P. 102: Bernoulli

P. 17: Mare Spumans

P. 29: Lacus Somniorum

P. 35 and P. 102: Bobillier

P. 39 and P. 107: Rabbi Levi

P. 39 and P. 102: Blanchinus

P. 62: Byrgius and Byrgius A

P. 67 and P. 105: Krafft

P. 69 and P. 104: Gerard

P. 73: Dziewulsky

P. 74 and P. 108: Scaliger

P. 75: Ganswindt

P. 76: Nasmyth

P. 76: Pilatre

P. 76: Wargentin

P. 77: Vasco da Gama

P. 78 and P. 109: Zsigmondy

P. 78 and P. 106: McLaughlin

P. 79 and P. 108: Schwabe

P. 96 and P. 100: Heaviside

P. 104: Grimaldi

P. 108: Smoky Mountains

Other errors

P. 73: Charts 9, 9, 6 are indicated at left – charts 2 and 3 should be indicated instead, see P. 71

P. 77 and P. 79: Nearside / Farside should be reversed

P. 94: Craters Collins and Aldrin are reversed in the first image

P. 95: Landing site of Luna 9 is on chart 27, not chart 26

P. 111: Volcanoes of North America was published in 1990, not 1900

Thanks to Danny Caes, Sally Russell, Frank McCabe, Howard Eskildsen, Dave Chapman, Rob Crow, Richard Mallett and Udo Schlegel for discovering these errors!